Data Visualization

Data Visualization

A PRACTICAL INTRODUCTION

Kieran Healy

PRINCETON UNIVERSITY PRESS

PRINCETON AND OXFORD

© 2019 Princeton University Press

Published by Princeton University Press
41 William Street, Princeton, New Jersey 08540
6 Oxford Street, Woodstock, Oxfordshire OX20 1TR

press.princeton.edu

All Rights Reserved

Library of Congress Control Number: 2018935810

ISBN 978-0-691-18161-5
ISBN (pbk.) 978-0-691-18162-2

British Library Cataloging-in-Publication Data is available

This book has been composed with open-source tools in Minion Pro,
Myriad Pro, and Iosevka Type.

Printed on acid-free paper. ∞

Printed in the United States of America

10 9 8 7 6 5 4

For the Llamanteriat, who saw it first.

Contents

Preface

You should look at your data. Graphs and charts let you explore and learn about the structure of the information you collect. Good data visualizations also make it easier to communicate your ideas and findings to other people. Beyond that, producing effective plots from your own data is the best way to develop a good eye for reading and understanding graphs—good and bad—made by others, whether presented in research articles, business slide decks, public policy advocacy, or media reports. This book teaches you how to do it.

My main goal is to introduce you to both the *ideas* and the *methods* of data visualization in a sensible, comprehensible, reproducible way. Some classic works on visualizing data, such as *The Visual Display of Quantitative Information* (Tufte 1983), present numerous examples of good and bad work together with some general taste-based rules of thumb for constructing and assessing graphs. In what has now become a large and thriving field of research, more recent work provides excellent discussions of the cognitive underpinnings of successful and unsuccessful graphics, again providing many compelling and illuminating examples (Ware 2008). Other books provide good advice about how to graph data under different circumstances (Cairo 2013; Few 2009; Munzer 2014) but choose not to teach the reader about the tools used to produce the graphics they show. This may be because the software used is some (proprietary, costly) point-and-click application that requires a fully visual introduction of its own, such as Tableau, Microsoft Excel, or SPSS. Or perhaps the necessary software is freely available, but showing how to use it is not what the book is about (Cleveland 1994). Conversely, there are excellent cookbooks that provide code "recipes" for many kinds of plot (Chang 2013). But for that reason they do not take the time to introduce the beginner to the principles behind the output they produce. Finally, we also have thorough introductions to particular software tools and libraries, including the ones we will use in this book (Wickham 2016). These can sometimes be hard for beginners to digest, as they may presuppose a background that the reader does not have.

Each of the books I have just cited is well worth your time. When teaching people how to make graphics with data, however, I have repeatedly found the need for an introduction that motivates and explains *why* you are doing something but that does not skip the necessary details of *how* to produce the images you see on the page. And so this book has two main aims. First, I want you to get to the point where you can reproduce almost every figure in the text for yourself. Second, I want you to understand why the code is written the way it is, such that when you look at data of your own you can feel confident about your ability to get from a rough picture in your head to a high-quality graphic on your screen or page.

What You Will Learn

This book is a hands-on introduction to the principles and practice of looking at and presenting data using R and ggplot. R is a powerful, widely used, and freely available programming language for data analysis. You may be interested in exploring ggplot after having used R before or be entirely new to both R and ggplot and just want to graph your data. I do not assume you have any prior knowledge of R.

After installing the software we need, we begin with an overview of some basic principles of visualization. We focus not just on the aesthetic aspects of good plots but on how their effectiveness is rooted in the way we perceive properties like length, absolute and relative size, orientation, shape, and color. We then learn how to produce and refine plots using ggplot2, a powerful, versatile, and widely used visualization package for R (Wickham 2016). The ggplot2 library implements a "grammar of graphics" (Wilkinson 2005). This approach gives us a coherent way to produce visualizations by expressing relationships between the attributes of data and their graphical representation.

Through a series of worked examples, you will learn how to build plots piece by piece, beginning with scatterplots and summaries of single variables, then moving on to more complex graphics. Topics covered include plotting continuous and categorical variables; layering information on graphics; faceting grouped data to produce effective "small multiple" plots; transforming data to easily produce visual summaries on the graph such as trend lines, linear fits, error ranges, and boxplots; creating maps; and some alternatives to maps worth considering when presenting country- or state-level data. We will also cover cases where we are not working directly with a dataset but rather with estimates from a statistical model. From there, we will explore the process of refining plots to accomplish common tasks such as highlighting key features of the data, labeling particular items of interest, annotating plots, and changing their overall appearance. Finally we will examine some strategies for presenting graphical results in different formats and to different sorts of audiences.

If you follow the text and examples in this book, then by the end you will

- understand the basic principles behind effective data visualization;
- have a practical sense for why some graphs and figures work well, while others may fail to inform or actively mislead;
- know how to create a wide range of plots in R using ggplot2; and
- know how to refine plots for effective presentation.

Learning how to visualize data effectively is more than just knowing how to write code that produces figures from data. This book will teach you how to do that. But it will also teach you how to think about the information you want to show, and how to consider the audience you are showing it to—including the most common case, when the audience is yourself.

This book is not a comprehensive guide to R, or even a comprehensive survey of everything ggplot can do. Nor is it a cookbook containing just examples of specific things people commonly want to do with ggplot. (Both these sorts of books already exist: see the references in the appendix.) Neither is it a rigid set of rules, or a sequence of beautiful finished examples that you can admire but not reproduce. My goal is to get you quickly up and running in R, making plots in a well-informed way, with a solid grasp of the core sequence of steps—taking your data, specifying the relationship between variables and visible elements, and building up images layer by layer—that is at the heart of what ggplot does.

Learning ggplot does mean getting used to how R works, and also understanding how ggplot connects to other tools in the R language. As you work your way through the book, you will gradually learn more about some very useful idioms, functions, and techniques for manipulating data in R. In particular you will learn about some of the tools provided by the tidyverse library that ggplot belongs to. Similarly, although this is not a cookbook, once you get past chapter 1 you will be able to see and understand the code used to produce almost every figure in the book. In most cases you will also see these figures built up piece by piece, a step at a time. If you use the book as it is designed, by the end you will have the makings of a version of the book itself, containing code you have written out and annotated yourself. And though we do not go into great depth on the topic of rules or principles of visualization, the discussion in chapter 1 and its application throughout the book gives you more to think about than just a list of graph types. By the end of the book you should be able to look at a figure and be able to see it in terms of ggplot's grammar, understanding how the various layers, shapes, and data are pieced together to make a finished plot.

The Right Frame of Mind

It can be a little disorienting to learn a programming language like R, mostly because at the beginning there seem to be so many pieces to fit together in order for things to work properly. It can seem like you have to learn everything before you can do anything. The language has some possibly unfamiliar concepts that define how it works, like "object," "function," or "class." The syntactic rules for writing code are annoyingly picky. Error messages seem obscure; help pages are terse; other people seem to have had *not quite* the same issue as you. Beyond that, you sense that doing one thing often involves learning a bit about some other part of the language. To make a plot you need a table of data, but maybe you need to filter out some rows, recalculate some columns, or just get the computer to see it is there in the first place. And there is also a wider environment of supporting applications and tools that are good to know about but involve new concepts of their own—editors that highlight what you write; applications that help you organize your code and its output; ways of writing your code that let you keep track of what you have done. It can all seem a bit confusing.

Don't panic. You have to start somewhere. Starting with graphics is more rewarding than some of the other places you might begin, because you will be able to see the results of your efforts very quickly. As you build your confidence and ability in this area, you will gradually see the other tools as things that help you sort out some issue or solve a problem that's stopping you from making the picture you want. That makes them easier to learn. As you acquire them piecemeal—perhaps initially using them without completely understanding what is happening—you will begin to see how they fit together and be more confident of your own ability to do what you need to do.

Even better, in the past decade or so the world of data analysis and programming generally has opened up in a way that has made help much easier to come by. Free tools for coding have been around for a long time, but in recent years what we might call the "ecology of assistance" has gotten better. There are more resources available for learning the various pieces, and more of them are oriented to the way writing code actually happens most of the time—which is to say, iteratively, in an error-prone fashion, and taking account of problems other people have run into and solved before.

How to Use This Book

This book can be used in any one of several ways. At a minimum, you can sit down and read it for a general overview of good practices in data visualization, together with many worked examples of graphics from their beginnings to a properly finished state. Even if you do not work through the code, you will get a good sense of how to think about visualization and a better understanding of the process through which good graphics are produced.

More useful, if you set things up as described in chapter 2 and then work through the examples, you will end up with a data visualization book of your own. If you approach the book this way, then by the end you will be comfortable using ggplot in particular and also be ready to learn more about the R language in general.

This book can also be used to teach with, either as the main focus of a course on data visualization or as a supplement to

You can also bring your own data to explore instead of or alongside the examples, as described in chapter 2.

undergraduate or graduate courses in statistics or data analysis. My aim has been to make the "hidden tasks" of coding and polishing graphs more accessible and explicit. I want to make sure you are not left with the "How to Draw an Owl in Three Steps" problem common to many tutorials. You know the one. The first two steps are shown clearly enough. Sketch a few bird-shaped ovals. Make a line for a branch. But the final step, an owl such as John James Audubon might have drawn, is presented as a simple extension for readers to figure out for themselves.

If you have never used R or ggplot, you should start at the beginning of the book and work your way through to the end. If you know about R already and only want to learn the core of ggplot, then after installing the software described below, focus on chapters 3 through 5. Chapter 6 (on models) necessarily incorporates some material on statistical modeling that the book cannot develop fully. This is not a statistics text. So, for example, I show generally how to fit and work with various kinds of model in chapter 6, but I do not go through the important details of fitting, selecting, and fully understanding different approaches. I provide references in the text to other books that have this material as their main focus.

Each chapter ends with a section suggesting where to go next (apart from continuing to read the book). Sometimes I suggest other books or websites to explore. I also ask questions or pose some challenges that extend the material covered in the chapter, encouraging you to use the concepts and skills you have learned.

Conventions

In this book we alternate between regular text (like this), samples of code that you can type and run yourself, and the output of that code. In the main text, references to objects or other things that exist in the R language or in your R project—tables of data, variables, functions, and so on—will also appear in a monospaced or "typewriter" typeface. Code you can type directly into R at the console will be in gray boxes and also monospaced, like this:

```
my_numbers ← c(1, 1, 4, 1, 1, 4, 1)
```

Additional notes and information will sometimes appear in the margin, like this.

If you type that line of code into R's console, it will create a thing called my_numbers. Doing this doesn't produce any output, however. When we write code that also produces output at the console, we will first see the code (in a gray box) and then the output in a monospaced font against a white background. Here we add two numbers and see the result:

```
4 + 1
```

```
## [1] 5
```

Two further notes about how to read this. First, by default in this book, anything that comes back to us at the console as the result of typing a command will be shown prefaced by two hash characters (##) at the beginning of each line of output. This is to help distinguish it from commands we type into the console. You will not see the hash characters at the console when you use R.

Second, both in the book *and* at the console, if the output of what you did results in a series of elements (numbers, observations from a variable, and so on), you will often see output that includes some number in square brackets at the beginning of the line. It looks like this: [1]. This is not part of the output itself but just a counter or index keeping track of how many items have been printed out so far. In the case of adding 4 + 1 we got just one, or [1], thing back—the number five. If there are more elements returned as the result of some instruction or command, the counter will keep track of that on each line. In this next bit of code we will tell R to show us the lowercase letters of the alphabet:

```
letters
```

```
##  [1] "a" "b" "c" "d" "e" "f" "g" "h" "i" "j"
## [11] "k" "l" "m" "n" "o" "p" "q" "r" "s" "t"
## [21] "u" "v" "w" "x" "y" "z"
```

You can see the counter incrementing on each line as it keeps count of how many letters have been printed.

Before You Begin

The book is designed for you to follow along in an active way, writing out the examples and experimenting with the code as you go. You will be able to reproduce almost all the plots in the text. You need to install some software first. Here is what to do:

1. Get the most recent version of R. It is free and available for Windows, Mac, and Linux operating systems. Download the version of R compatible with your operating system. If you are running Windows or MacOS, choose one of the *precompiled binary* distributions (i.e., ready-to-run applications) linked at the top of the R Project's web page.

2. Once R is installed, download and install R Studio, which is an "Integrated Development Environment," or IDE. This means it is a front-end for R that makes it much easier to work with. R Studio is also free and available for Windows, Mac, and Linux platforms.

3. Install the tidyverse and several other add-on packages for R. These packages provide useful functionality that we will take advantage of throughout the book. You can learn more about the tidyverse's family of packages at its website.

To install the tidyverse, make sure you have an internet connection and then launch R Studio. Type the following lines of code at R's command prompt, located in the window named "Console," and hit return. In the code below, the ← arrow is made up of two keystrokes, first < and then the short dash or minus symbol, -.

I strongly recommend typing all the code examples right from the beginning, instead of copying and pasting.

```
my_packages ← c("tidyverse", "broom", "coefplot", "cowplot",
                "gapminder", "GGally", "ggrepel", "ggridges", "gridExtra",
                "here", "interplot", "margins", "maps", "mapproj",
                "mapdata", "MASS", "quantreg", "rlang", "scales",
                "survey", "srvyr", "viridis", "viridisLite", "devtools")

install.packages(my_packages, repos = "http://cran.rstudio.com")
```

R Studio should then download and install these packages for you. It may take a little while to download everything.

With these packages available, you can then install one last library of material that's useful specifically for this book. It is hosted on GitHub, rather than R's central package repository, so we use a different function to fetch it.

github.com

GitHub is a web-based service where users can host, develop, and share code. It uses git, a version control system that allows projects, or repositories, to preserve their history and incorporate changes from contributors in an organized way.

```
devtools::install_github("kjhealy/socviz")
```

Once you've done that, we can get started.

Data Visualization

1   Look at Data

Some data visualizations are better than others. This chapter discusses why that is. While it is tempting to simply start laying down the law about what works and what doesn't, the process of making a really good or really useful graph cannot be boiled down to a list of simple rules to be followed without exception in all circumstances. The graphs you make are meant to be looked at by someone. The effectiveness of any particular graph is not just a matter of how it looks in the abstract but also a question of who is looking at it, and why. An image intended for an audience of experts reading a professional journal may not be readily interpretable by the general public. A quick visualization of a dataset you are currently exploring might not be of much use to your peers or your students.

Some graphs work well because they depend in part on some strong aesthetic judgments about what will be effective. That sort of good judgment is hard to systematize. However, data visualization is not simply a matter of competing standards of good taste. Some approaches work better for reasons that have less to do with one's sense of what looks good and more to do with how human visual perception works. When starting out, it is easier to grasp these perceptual aspects of data visualization than it is to get a reliable, taste-based feel for what works. For this reason, it is better to begin by thinking about the relationship between the structure of your data and the perceptual features of your graphics. Getting into that habit will carry you a long way toward developing the ability to make good taste-based judgments, too.

As we shall see later on, when working with data in R and ggplot, we get many visualization virtues for free. In general, the default layout and look of ggplot's graphics is well chosen. This makes it easier to do the right thing. It also means that, if you *really* just want to learn how to make some plots right this minute, you could skip this chapter altogether and go straight to the next one. But although we will not be writing any code for the next few pages, we will be discussing aspects of graph construction, perception, and interpretation that matter for code you will choose to write. So

I urge you to stick around and follow the argument of this chapter. When making graphs there is only so much that your software can do to keep you on the right track. It cannot force you to be honest with yourself, your data, and your audience. The tools you use can help you live up to the right standards. But they cannot make you do the right thing. This means it makes sense to begin cultivating your own good sense about graphs right away.

We will begin by asking why we should bother to look at pictures of data in the first place, instead of relying on tables or numerical summaries. Then we will discuss a few examples, first of bad visualization practice, and then more positively of work that looks (and is) much better. We will examine the usefulness and limits of general rules of thumb in visualization and show how even tasteful, well-constructed graphics can mislead us. From there we will briefly examine some of what we know about the perception of shapes, colors, and relationships between objects. The core point here is that we are quite literally able to see some things much more easily than others. These cognitive aspects of data visualization make some kinds of graphs reliably harder for people to interpret. Cognition and perception are relevant in other ways, too. We tend to make inferences about relationships between the objects that we see in ways that bear on our interpretation of graphical data, for example. Arrangements of points and lines on a page can encourage us—sometimes quite unconsciously—to make inferences about similarities, clustering, distinctions, and causal relationships that might or might not be there in the numbers. Sometimes these perceptual tendencies can be honestly harnessed to make our graphics more effective. At other times, they will tend to lead us astray, and we must take care not to lean on them too much.

In short, good visualization methods offer extremely valuable tools that we should use in the process of exploring, understanding, and explaining data. But they are not a magical means of seeing the world as it really is. They will not stop you from trying to fool other people if that is what you want to do, and they may not stop you from fooling yourself either.

1.1 Why Look at Data?

Anscombe's quartet (Anscombe 1973; Chatterjee & Firat 2007), shown in figure 1.1, presents its argument for looking at data in visual form. It uses a series of four *scatterplots*. A scatterplot shows

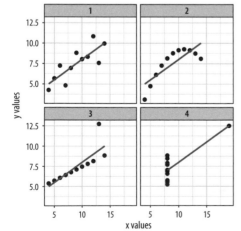

Figure 1.1: Plots of Anscombe's quartet.

the relationship between two quantities, such as height and weight, age and income, or time and unemployment. Scatterplots are the workhorse of data visualization in social science, and we will be looking at a lot of them. The data for Anscombe's plots comes bundled with R. You can look at it by typing anscombe at the command prompt. Each of the four made-up "datasets" contains eleven observations of two variables, x and y. By construction, the numerical properties of each pair of x and y variables, such as their means, are almost identical. Moreover, the standard measures of the association between each x and y pair also match. The correlation coefficient is a strong 0.81 in every case. But when the datasets are visualized as a scatterplot, with the x variables plotted on the horizontal axis and the y variables on the vertical, the differences between them are readily apparent.

Anscombe's quartet is an extreme, manufactured example. But the benefits of visualizing one's data can be shown in real cases. Figure 1.2 shows a graph from Jackman (1980), a short comment on Hewitt (1977). The original paper had argued for a significant association between voter turnout and income inequality based on a quantitative analysis of eighteen countries. When this relationship was graphed as a scatterplot, however, it immediately became clear that the quantitative association depended entirely on the inclusion of South Africa in the sample.

An exercise by Jan Vanhove (2016) demonstrates the usefulness of looking at model fits and data at the same time. Figure 1.3 presents an array of scatterplots. As with Anscombe's quartet, each panel shows the association between two variables. Within each panel, the correlation between the x and y variables is set to be 0.6, a pretty good degree of association. But the actual distribution of points is created by a different process in each case. In the top left panel each variable is normally distributed around its mean value. In other panels there is a single outlying point far off in one direction or another. Others are produced by more subtle rules. But each gives rise to the same basic linear association.

Illustrations like this demonstrate why it is worth looking at data. But that does not mean that looking at data is all one needs to do. Real datasets are messy, and while displaying them graphically is very useful, doing so presents problems of its own. As we will see below, there is considerable debate about what sort of visual work is most effective, when it can be superfluous, and how it can at times be misleading to researchers and audiences alike. Just like seemingly sober and authoritative tables of numbers, data

Correlations can run from -1 to 1, with zero meaning there is no association. A score of -1 means a perfect negative association and a score of 1 a perfect positive asssociation between the two variables. So 0.81 counts as a strong positive correlation.

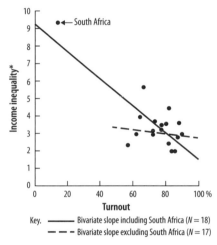

Figure 1.2: Seeing the effect of an outlier on a regression line.

A more careful quantitative approach could have found this issue as well, for example, with a proper sensitivity analysis. But the graphic makes the case directly.

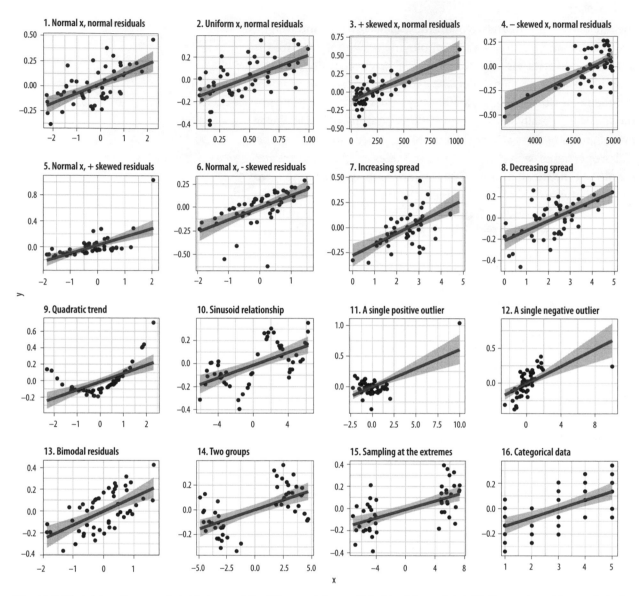

Figure 1.3: What data patterns can lie behind a correlation? The correlation coefficient in all these plots is 0.6. Figure adapted from code by Jan Vanhove.

visualizations have their own rhetoric of plausibility. Anscombe's quartet notwithstanding, and especially for large volumes of data, summary statistics and model estimates should be thought of as tools that we use to *deliberately* simplify things in a way that lets us see *past* a cloud of data points shown in a figure. We will not automatically get the right answer to our questions just by looking.

1.2 What Makes Bad Figures Bad?

It is traditional to begin discussions of data visualization with a "parade of horribles," in an effort to motivate good behavior later. However, these negative examples often combine several kinds of badness that are better kept separate. For convenience, we can say that our problems tend to come in three varieties. Some are strictly *aesthetic*. The graph we are looking at is in some way tacky, tasteless, or a hodgepodge of ugly or inconsistent design choices. Some are *substantive*. Here, our graph has problems that are due to the data being presented. Good taste might make things look better, but what we really need is to make better use of the data we have, or get new information and plot that instead. And some problems are *perceptual*. In these cases, even with good aesthetic qualities and good data, the graph will be confusing or misleading because of how people perceive and process what they are looking at. It is important to understand that these elements, while often found together, are distinct from one another.

Bad taste

Let's start with the bad taste. The chart in figure 1.4 both is tasteless and has far too much going on in it, given the modest amount of information it displays. The bars are hard to read and compare. It needlessly duplicates labels and makes pointless use of three-dimensional effects, drop shadows, and other unnecessary design features.

The best-known critic by far of this style of visualization, and the best-known taste-maker in the field, is Edward R. Tufte. His book *The Visual Display of Quantitative Information* (1983) is a classic, and its sequels are also widely read (Tufte 1990, 1997). The

Figure 1.4: A chart with a considerable amount of junk in it.

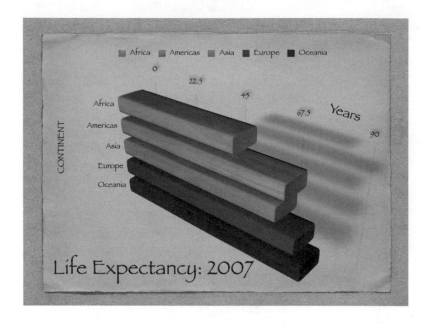

bulk of this work is a series of examples of good and bad visualization, along with some articulation of more general principles (or rules of thumb) extracted from them. It is more like a reference book about completed dishes than a cookbook for daily use in the kitchen. At the same time, Tufte's early academic work in political science shows that he effectively applied his own ideas to research questions. His *Political Control of the Economy* (1978) combines tables, figures, and text in a manner that remains remarkably fresh almost forty years later.

Tufte's message is sometimes frustrating, but it is consistent:

> Graphical excellence is the well-designed presentation of interesting data—a matter of substance, of statistics, and of design.... [It] consists of complex ideas communicated with clarity, precision, and efficiency.... [It] is that which gives to the viewer the greatest number of ideas in the shortest time with the least ink in the smallest space.... [It] is nearly always multivariate.... And graphical excellence requires telling the truth about the data. (Tufte 1983, 51)

Tufte illustrates the point with Charles Joseph Minard's famous visualization of Napoleon's march on Moscow, shown here in

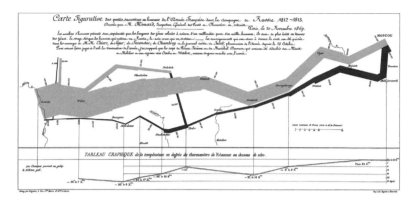

Figure 1.5: Minard's visualization of Napoleon's retreat from Moscow. Justifiably cited as a classic, it is also atypical and hard to emulate in its specifics.

figure 1.5. He remarks that this image "may well be the best statistical graphic ever drawn" and argues that it "tells a rich, coherent story with its multivariate data, far more enlightening than just a single number bouncing along over time. Six variables are plotted: the size of the army, its location on a two-dimensional surface, direction of the army's movement, and temperature on various dates during the retreat from Moscow."

It is worth noting how far removed Minard's image is from most contemporary statistical graphics. At least until recently, these have tended to be applications or generalizations of scatterplots and bar plots, in the direction of either seeing more raw data or seeing the output derived from a statistical model. The former looks for ways to increase the volume of data visible, or the number of variables displayed within a panel, or the number of panels displayed within a plot. The latter looks for ways to see results such as point estimates, confidence intervals, and predicted probabilities in an easily comprehensible way. Tufte acknowledges that a tour de force such as Minard's "can be described and admired, but there are no compositional principles on how to create that one wonderful graphic in a million." The best one can do for "more routine, workaday designs" is to suggest some guidelines such as "have a properly chosen format and design," "use words, numbers, and drawing together," "display an accessible complexity of detail," and "avoid content-free decoration, including chartjunk" (Tufte 1983, 177).

In practice those compositional principles have amounted to an encouragement to maximize the "data-to-ink" ratio. This is practical advice. It is not hard to jettison tasteless junk, and if

Figure 1.6: "Monstrous Costs" by Nigel Holmes (1982). Also a classic of its kind.

we look a little harder we may find that the chart can do without other visual scaffolding as well. We can often clean up the typeface, remove extraneous colors and backgrounds, and simplify, mute, or delete gridlines, superfluous axis marks, or needless keys and legends. Given all that, we might think that a solid rule of "simpify, simplify" is almost all of what we need to make sure that our charts remain junk-free and thus effective. Unfortunately this is not the case. For one thing, somewhat annoyingly, there is evidence that highly embellished charts like Nigel Holmes's "Monstrous Costs" (fig. 1.6) are often more easily recalled than their plainer alternatives (Bateman et al. 2010). Viewers do not find them more easily interpretable, but they do remember them more easily and also seem to find them more enjoyable to look at. They also associate them more directly with value judgments, as opposed to just trying to get information across. Borkin et al. (2013) also found that visually unique, "infographic"-style graphs were more memorable than more standard statistical visualizations. ("It appears that novel and unexpected visualizations can be better remembered than the visualizations with limited variability that we are exposed to since elementary school," they remark.)

Even worse, it may be the case that graphics that really do maximize the data-to-ink ratio are harder to interpret than those that are a little more relaxed about it. E. W. Anderson et al. (2011) found that, of the six kinds of boxplot shown in figure 1.7, the minimalist version from Tufte's own work (option C) proved to be the most cognitively difficult for viewers to interpret. Cues like labels and gridlines, together with some strictly superfluous embellishment of data points or other design elements, may often be an aid rather than an impediment to interpretation.

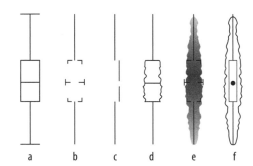

Figure 1.7: Six kinds of summary boxplots. Type (c) is from Tufte.

While chartjunk is not entirely devoid of merit, bear in mind that ease of recall is only one virtue among many for graphics. It is also the case that, almost by definition, it is no easier to systematize the construction of a chart like "Monstrous Costs" than it is to replicate the impact of Minard's graph of Napoleon's retreat. Indeed, the literature on chartjunk suggests that the two may have some qualities in common. To be sure, Minard's figure is admirably rich in data while Holmes's is not. But both are visually distinctive in a way that makes them memorable, both show a substantial amount of bespoke design, and both are unlike most of the statistical graphs you will see or make.

Bad data

In your everyday work you will be in little danger of producing either a "Monstrous Costs" or a "Napoleon's Retreat." You are much more likely to make a good-looking, well-designed figure that misleads people because you have used it to display some bad data. Well-designed figures with little or no junk in their component parts are not by themselves a defense against cherry-picking your data or presenting information in a misleading way. Indeed, it is even possible that, in a world where people are on guard against junky infographics, the "halo effect" accompanying a well-produced figure might make it *easier* to mislead some audiences. Or, perhaps more common, good aesthetics does not make it much harder for you to mislead yourself as you look at your data.

In November 2016 the *New York Times* reported on some research on people's confidence in the institutions of democracy. It had been published in an academic journal by the political scientists Yascha Mounk and Roberto Stefan Foa. The headline in the *Times* ran "How Stable Are Democracies? 'Warning Signs Are Flashing Red' " (Taub 2016). The graph accompanying the article, reproduced in figure 1.8, certainly seemed to show an alarming decline.

The graph was widely circulated on social media. It is impressively well produced. It's an elegant small-multiple that, in addition to the point ranges it identifies, also shows an error range (labeled as such for people who might not know what it is), and the story told across the panels for each country is pretty consistent.

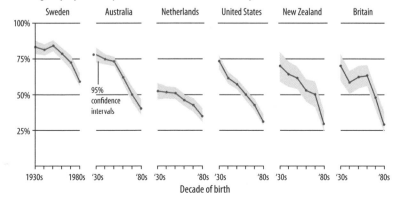

Figure 1.8: A crisis of faith in democracy? (Source: Roberto Stefan Foa and Yascha Mounk, "The Signs of Deconsolidation," *Journal of Democracy*, *28*(1), 5–16.)

The figure is a little tricky to interpret. As the x-axis label says, the underlying data are from a cross-sectional survey of people of different ages rather than a longitudinal study measuring everyone at different times. Thus the lines do not show a trend measured each decade from the 1930s but rather differences in the answers given by people born in different decades, all of whom were asked the question at the same time. Given that, a bar graph might have been a more appropriate to display the results.

More important, as the story circulated, helped by the compelling graphic, scholars who knew the World Values Survey data underlying the graph noticed something else. The graph reads as though people were asked to say whether they thought it was essential to live in a democracy, and the results plotted show the percentage of respondents who said "Yes," presumably in contrast to those who said "No." But in fact the survey question asked respondents to rate the importance of living in a democracy on a ten-point scale, with 1 being "Not at all Important" and 10 being "Absolutely Important." The graph showed the difference across ages of people who had given a score of 10 only, not changes in the average score on the question. As it turns out, while there is some variation by year of birth, most people in these countries tend to rate the importance of living in a democracy very highly, even if they do not all score it as "Absolutely Important." The political scientist Erik Voeten redrew the figure using the average response. The results are shown in figure 1.9.

One reason I chose this example is that, at the time of writing, it is not unreasonable to be concerned about the stability of people's commitment to democratic government in some Western countries. Perhaps Mounk's argument is correct. But in such cases, the question is how much we are letting the data speak to us, as opposed to arranging it to say what we already think for other reasons.

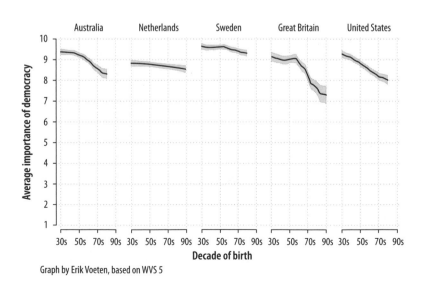

Figure 1.9: Perhaps the crisis has been overblown. (Erik Voeten.)

The change here is *not* due to a difference in how the y-axis is drawn. That is a common issue with graphs, and one we will discuss below. In this case both the *New York Times* graph and Voeten's alternative have scales that cover the full range of possible values (from 0 to 100% in the former case and from 1 to 10 in the latter). Rather, a different measure is being shown. We are now looking at the trend in the average score, rather than the trend for the highest possible answer. Substantively, there *does* still seem to be a decline in the average score by age cohort, on the order of between 0.5 point and 1.5 points on a 10-point scale. It could be an early warning sign of a collapse of belief in democracy, or it could be explained by something else. It might even be reasonable (as we will see for a different example shortly) to present the data in Voeten's version with the y-axis covering just the range of the decline, rather than the full 0–10 scale. But it seems fair to say that the story might not have made the *New York Times* if the original research article had presented Voeten's version of the data rather than the one that appeared in the newspaper.

Bad perception

Our third category of badness lives in the gap between data and aesthetics. Visualizations encode numbers in lines, shapes, and colors. That means that our interpretation of these encodings is partly conditional on how we perceive geometric shapes and relationships generally. We have known for a long time that poorly encoded data can be misleading. Tufte (1983) contains many examples, as does Wainer (1984). Many of the instances they cite revolve around needlessly multiplying the number of dimensions shown in a plot. Using an area to represent a length, for example, can make differences between observations look larger than they are.

Although the most egregious abuses are less common than they once were, adding additional dimensions to plots remains a common temptation. Figure 1.10, for instance, is a 3-D bar chart made using a recent version of Microsoft Excel. Charts like this are common in business presentations and popular journalism and are also seen in academic journal articles from time to time. Here we seek to avoid too much junk by using Excel's default settings. As

To be fair, the 3-D format is not Excel's default type of bar chart.

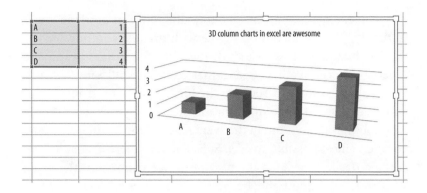

Figure 1.10: A 3-D column chart created in Microsoft Excel for Mac. Although it may seem hard to believe, the values shown in the bars are 1, 2, 3, and 4.

you can see from the cells shown to the left of the chart, the data we are trying to plot is not very complex. The chart even tries to help us by drawing and labeling grid lines on the y- (and z-) axes. And yet the 3-D columns in combination with the default angle of view for the chart make the values as displayed differ substantially from the ones actually in the cell. Each column appears to be somewhat below its actual value. It is possible to see, if you squint with your mind's eye, how the columns would line up with the axis guidelines if your angle of view moved so that the bars were head-on. But as it stands, anyone asked what values the chart shows would give the wrong answer.

By now, many regular users of statistical graphics know enough to avoid excessive decorative embellishments of charts. They are also usually put on their guard by overly elaborate presentation of simple trends, as when a three-dimensional ribbon is used to display a simple line. Moreover, the default settings of most current graphical software tend to make the user work a little harder to add these features to plots.

Even when the underlying numbers are sensible, the default settings of software are good, and the presentation of charts is mostly junk-free, some charts remain more difficult to interpret than others. They encode data in ways that are hard for viewers to understand. Figure 1.11 presents a stacked bar chart with time in years on the x-axis and some value on the y-axis. The bars show the total value, with subdivisions by the relative contribution of different categories to each year's observation. Charts like this are common when showing the absolute contribution of various products to total sales over time, for example, or the number of different groups of people in a changing population. Equivalently, stacked

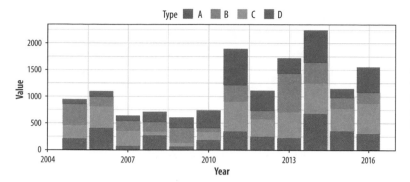

Figure 1.11: A junk-free plot that remains hard to interpret. While a stacked bar chart makes the overall trend clear, it can make it harder to see the trends for the categories within the bar. This is partly due to the nature of the trends. But if the additional data is hard to understand, perhaps it should not be included to begin with.

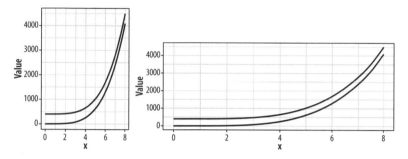

Figure 1.12: Aspect ratios affect our perception of rates of change. (After an example by William S. Cleveland.)

line-graphs showing similar kinds of trends are also common for data with many observation points on the x-axis, such as quarterly observations over a decade.

In a chart like this, the overall trend is readily interpretable, and it is also possible to easily follow the over-time pattern of the category that is closest to the x-axis baseline (in this case, type D, in purple). But the fortunes of the other categories are not so easily grasped. Comparisons of both the absolute and the relative share of type B or C are much more difficult, whether one wants to compare trends within type or between them. Relative comparisons need a stable baseline. In this case, that's the x-axis, which is why the overall trend and the type D trend are much easier to see than any other trend.

A different sort of problem is shown in figure 1.12. In the left panel, the lines appear at first glance to be converging as the value of x increases. It seems like they might even intersect if we extended the graph out further. In the right panel, the curves are clearly equidistant from the beginning. The data plotted in each panel is the same, however. The apparent convergence in the left panel is just a result of the aspect ratio of the figure.

These problems are not easily solved by the application of good taste, or by following a general rule to maximize the data-to-ink ratio, even though that is a good rule to follow. Instead, we need to know a little more about the role of perception in the interpretation of graphs. Fortunately for us, this is an area that has produced a substantial amount of research over the past twenty-five years.

1.3 Perception and Data Visualization

While a detailed discussion of visual perception is well beyond the scope of this book, even a simple sense of how we see things will help us understand why some figures work and others do not. For a much more thorough treatment of these topics, Colin Ware's books on information design are excellent overviews of research on visual perception, written from the perspective of people designing graphs, figures, and systems for representing data (Ware 2008, 2013).

Edges, contrasts, and colors

Looking at pictures of data means looking at lines, shapes, and colors. Our visual system works in a way that makes some things easier for us to see than others. I am speaking in slightly vague terms here because the underlying details are the remit of vision science, and the exact mechanisms responsible are often the subject of ongoing research. I will not pretend to summarize or evaluate this material. In any case, independent of detailed explanation, the existence of the perceptual phenomena themselves can often be directly demonstrated through visual effects or "optical illusions" of various kinds. These effects demonstrate that perception is not a simple matter of direct visual inputs producing straightforward mental representations of their content. Rather, our visual system is tuned to accomplish some tasks very well, and this comes at a cost in other ways.

The active nature of perception has long been recognized. The Hermann grid effect, shown in figure 1.13, was discovered in 1870. Ghostly blobs seem to appear at the intersections in the grid but only as long as one is not looking at them directly. A related effect is shown in figure 1.14. These are *Mach bands*. When the gray bars share a boundary, the apparent contrast between them appears to

Figure 1.13: Hermann grid effect.

Figure 1.14: Mach bands. On the left side, five gray bars are ordered from dark to light, with gaps between them. On the right side, the bars have no gap between them. The brightness or luminance of the corresponding bars is the same. However, when the bars touch, the dark areas seem darker and the light areas lighter.

increase. Speaking loosely, we can say that our visual system is trying to construct a representation of what it is looking at based more on *relative* differences in the luminance (or brightness) of the bars rather than on their absolute value. Similarly, the ghostly blobs in the Hermann grid effect can be thought of as a side-effect of the visual system being tuned for a different task.

These sorts of effects extend to the role of background contrasts. The same shade of gray will be perceived differently depending on whether it is against a dark background or a light one. Our ability to distinguish shades of brightness is not uniform either. We are better at distinguishing dark shades than we are at distinguishing light ones. The effects interact, too. We will do better at distinguishing very light shades of gray when they are set against a light background. When set against a dark background, differences in the middle range of the light-to-dark spectrum are easier to distinguish.

Our visual system is attracted to edges, and we assess contrast and brightness in terms of relative rather than absolute values. Some of the more spectacular visual effects exploit our mostly successful efforts to construct representations of surfaces, shapes, and objects based on what we are seeing. Edward Adelson's checkershadow illusion, shown in figure 1.15, is a good example. Though hard to believe, the squares marked "A" and "B" are the same shade of gray.

To figure out the shade of the squares on the floor, we compare it to the nearby squares, and we also discount the shadows cast by other objects. Even though a light-colored surface in shadow might reflect less light than a dark surface in direct light, it would generally be an error to infer that the surface in the shade really was a darker color. The checkerboard image is carefully constructed to exploit these visual inferences made based on local contrasts in brightness and the information provided by shadows. As Adelson (1995) notes, "The visual system is not very good at being a physical light meter, but that is not its purpose." Because it has

Figure 1.15: The checkershadow illusion (Edward H. Adelson).

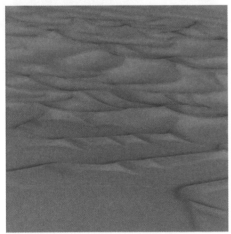

Figure 1.16: Edge contrasts in monochrome and color, after Ware (2008).

evolved to be good at perceiving real objects in its environment, we need to be aware of how it works in settings where we are using it to do other things, such as keying variables to some spectrum of grayscale values.

An important point about visual effects of this kind is that they are not illusions in the way that a magic trick is an illusion. If a magician takes you through an illusion step by step and shows you how it is accomplished, then the next time you watch the trick performed you will see through it and notice the various bits of misdirection and sleight of hand that are used to achieve the effect. But the most interesting visual effects are not like this. Even after they have been explained to you, you cannot stop seeing them, because the perceptual processes they exploit are not under your conscious control. This makes it easy to be misled by them, as when (for example) we overestimate the size of a contrast between two adjacent shaded areas on a map or grid simply because they share a boundary.

Our ability to see edge contrasts is stronger for monochrome images than for color. Figure 1.16, from Ware (2008, 71), shows an image of dunes. In the red-green version, the structure of the landscape is hard to perceive. In the grayscale version, the dunes and ridges are much more easily visible.

Using color in data visualization introduces a number of other complications (Zeileis & Hornik 2006). The central one is related to the relativity of luminance perception. As we have been discussing, our perception of how bright something looks is largely a matter of relative rather than absolute judgments. How bright a surface looks depends partly on the brightness of objects near it. In addition to luminance, the color of an object can be thought of has having two other components. First, an object's *hue* is what we conventionally mean when we use the word "color": red, blue, green, purple, and so on. In physical terms it can be thought of as the dominant wavelength of the light reflected from the object's surface. The second component is *chrominance* or *chroma*. This is the *intensity* or *vividness* of the color.

To produce color output on screens or in print we use various *color models* that mix together color components to get specific outputs. Using the RGB model, a computer might represent color in terms of mixtures of red, green, and blue components, each of which can take a range of values from 0 to 255. When using colors in a graph, we are mapping some quantity or category in our data to a color that people see. We want that mapping to be "accurate" in some sense, with respect to the data. This is partly a matter of the mapping being correct in strictly numerical terms. For instance, we want the gap between two numerical values in the data to be meaningfully preserved in the numerical values used to define the colors shown. But it is also partly a matter of how that mapping will be perceived when we look at the graph.

For example, imagine we had a variable that could take values from 0 to 5 in increments of 1, with zero being the lowest value. It is straightforward to map this variable to a set of RGB colors that are equally distant from one another in purely numerical terms in our color space. The wrinkle is that many points that are equidistant from each other in this sense will not be perceived as equally distant by people looking at the graph. This is because our perception is not uniform across the space of possible colors. For instance, the range of chroma we are able to see depends strongly on luminance. If we pick the wrong color palette to represent our data, for any particular gradient the same-sized jump between one value and another (e.g., from 0 to 1, as compared to from 3 to 4) might be perceived differently by the viewer. This also varies across colors, in that numerically equal gaps between a sequences of reds (say) are perceived differently from the same gaps mapped to blues.

When choosing color schemes, we will want mappings from data to color that are not just numerically but also *perceptually* uniform. R provides color models and color spaces that try to achieve this. Figure 1.17 shows a series of sequential gradients using the HCL (hue-chroma-luminance) color model. The grayscale gradient at the top varies by luminance only. The blue palette varies by luminance and chrominance, as the brightness and the intensity of the color vary across the spectrum. The remaining three palettes

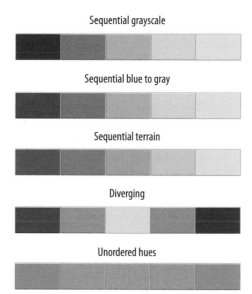

Sequential grayscale

Sequential blue to gray

Sequential terrain

Diverging

Unordered hues

Figure 1.17: Five palettes generated from R's color space library. From top to bottom, the sequential grayscale palette varies only in luminance, or brightness. The sequential blue palette varies in both luminance and chrominance (or intensity). The third sequential palette varies in luminance, chrominance, and hue. The fourth palette is diverging, with a neutral midpoint. The fifth features balanced hues, suitable for unordered categories.

vary by luminance, chrominance, and hue. The goal in each case is to generate a perceptually uniform scheme, where hops from one level to the next are seen as having the same magnitude.

Gradients or *sequential* scales from low to high are one of three sorts of color palettes. When we are representing a scale with a neutral midpoint (as when we are showing temperatures, for instance, or variance in either direction from a zero point or a mean value), we want a *diverging* scale, where the steps away from the midpoint are perceptually even in both directions. The blue-to-red palette in figure 1.17 displays an example. Finally, perceptual uniformity matters for unordered categorical variables as well. We often use color to represent data for different countries, or political parties, or types of people, and so on. In those cases we want the colors in our *qualitative* palette to be easily distinguishable but also have the same valence for the viewer. Unless we are doing it deliberately, we do not want one color to perceptually dominate the others. The bottom palette in figure 1.17 shows an example of a qualitative palette that is perceptually uniform in this way.

The upshot is that we should generally not pick colors in an ad hoc way. It is too easy to go astray. In addition to the considerations we have been discussing, we also want to avoid producing plots that confuse people who are color-blind, for example. Fortunately, almost all the work has been done for us already. Different color spaces have been defined and standardized in ways that account for these uneven or nonlinear aspects of human color perception. R and ggplot make these features available to us for free. The default palettes we will be using in ggplot are perceptually uniform in the right way. If we want to get more adventurous later, the tools are available to produce custom palettes that still have desirable perceptual qualities. Our decisions about color will focus more on when and how it should be used. As we are about to see, color is a powerful channel for picking out visual elements of interest.

Preattentive search and what "pops"

Some objects in our visual field are easier to see than others. They pop out at us from whatever they are surrounded by. For some kinds of object, or through particular channels, this can happen very quickly. Indeed, from our point of view it happens before or

The body responsible for this is the appropriately authoritative-sounding Commission Internationale de l'Eclairage, or International Commission on Illumination.

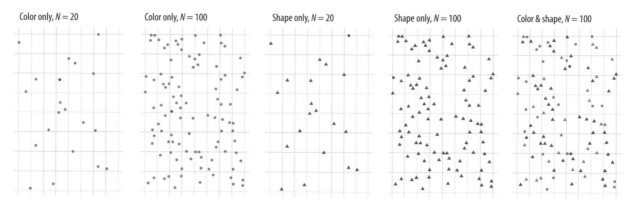

Figure 1.18: Searching for the blue circle becomes progressively harder.

almost before the conscious act of looking at or for something. The general term for this is "preattentive pop-out," and there is an extensive experimental and theoretical literature on it in psychology and vision science. As with the other perceptual processes we have been discussing, the explanation for what is happening is or has been a matter of debate, up to and including the degree to which the phenomenon really is "preattentive," as discussed, for example, by Treisman & Gormican (1988) or Nakayama & Joseph (1998). But it is the existence of pop-out that is relevant to us, rather than its explanation. Pop-out makes some things on a data graphic easier to see or find than others.

Consider the panels in figure 1.18. Each one of them contains a single blue circle. Think of it as an observation of interest. Reading left to right, the first panel contains twenty circles, nineteen of which are yellow and one blue. The blue circle is easy to find, as there are a relatively small number of observations to scan, and their color is the only thing that varies. The viewer barely has to search consciously at all before seeing the dot of interest.

In the second panel, the search is harder, but not that much harder. There are a hundred dots now, five times as many, but again the blue dot is easily found. The third panel again has only twenty observations. But this time there is no variation on color. Instead nineteen observations are triangles and one is a circle. On average, looking for the blue dot is noticeably harder than searching for it in the first panel, and it may even be more difficult than in the second panel despite there being many fewer observations.

Think of shape and color as two distinct *channels* that can be used to encode information visually. It seems that pop-out on the color channel is stronger than it is on the shape channel. In the fourth panel, the number of observations is again upped to one hundred. Finding the single blue dot may take noticeably longer. If you don't see it on the first or second pass, it may require a conscious effort to systematically scan the area in order to find it. It seems that search performance on the shape channel degrades much faster than on the color channel.

Finally the fifth panel mixes color and shape for a large number of observations. Again there is only one blue dot on the graph, but annoyingly there are many blue triangles and yellow dots that make it harder to find what we are looking for. Dual- or multiple-channel searches for large numbers of observations can be very slow.

Similar effects can be demonstrated for search across other channels (for instance, with size, angle, elongation, and movement) and for particular kinds of searches within channels. For example, some kinds of angle contrasts are easier to see than others, as are some kinds of color contrasts. Ware (2008, 27–33) has more discussion and examples. The consequences for data visualization are clear enough. As shown in figure 1.19, adding multiple channels to a graph is likely to quickly overtax the capacity of the viewer. Even if our software allows us to, we should think carefully before representing different variables and their values by shape, color, and position all at once. It is possible for there to be exceptions, in particular (as shown in the second panel of

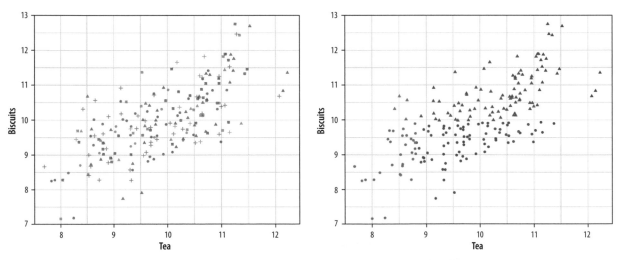

Figure 1.19: Multiple channels become uninterpretable very fast (*left*), unless your data has a great deal of structure (*right*).

figure 1.19) if the data shows a great deal of structure to begin with. But even here, in all but the most straightforward cases a different visualization strategy is likely to do better.

Gestalt rules

At first glance, the points in the pop-out examples in figure 1.18 might seem randomly distributed within each panel. In fact, they are not quite randomly located. Instead, I wrote a little code to lay them out in a way that spread them around the plotting area but prevented any two points from completely or partially overlapping each other. I did this because I wanted the scatterplots to be programmatically generated but did not want to take the risk that the blue dot would end up plotted underneath one of the other dots or triangles. It's worth taking a closer look at this case, as there is a lesson here for how we perceive patterns.

Each panel in figure 1.20 shows a field of points. There are clearly differences in structure between them. The first panel was produced by a two-dimensional Poisson point process and is "properly" random. (Defining randomness, or ensuring that a process really is random, turns out to be a lot harder than you might think. But we gloss over those difficulties here.) The second panel was produced from a Matérn model, a specification often found in spatial statistics and ecology. In a model like this points are again randomly distributed but are subject to some local constraints. In this case, after randomly generating a number of candidate points in order, the field is pruned to eliminate any point that appears too close to a point that was generated before it. We can tune the model to decide how close is "too close." The result is a set of points that are evenly spread across the available space.

If you ask people which of these panels has more structure in it, they will tend to say the Poisson field. We associate randomness with a relatively even distribution across a space. But in fact, a random process like this is substantially more clumpy than we tend to think. I first saw a picture of this contrast in an essay by Stephen Jay Gould (1991). There the Matérn-like model was used as a representation of glowworms on the wall of a cave in New Zealand. It's a good model for that case because if one glowworm gets too close to another, it's liable to get eaten. Hence the relatively even—but not random—distribution that results.

Poisson

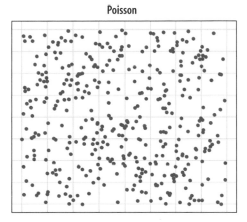

Matérn

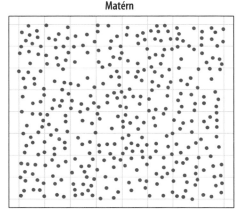

Figure 1.20: Each panel shows simulated data. The upper panel shows a random point pattern generated by a Poisson process. The lower panel is from a Matérn model, where new points are randomly placed but cannot be too near already-existing ones. Most people see the Poisson-generated pattern as having more structure, or less "randomness," than the Matérn, whereas the reverse is true.

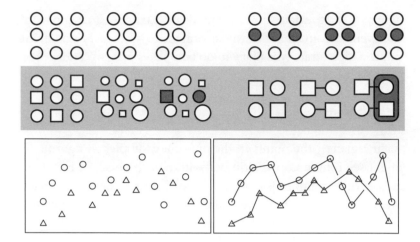

Figure 1.21: Gestalt inferences: proximity, similarity, connection, common fate. The layout of the figure employs some of these principles, in addition to displaying them.

We look for structure all the time. We are so good at it that we will find it in random data, given time. (This is one of the reasons that data visualization can hardly be a replacement for statistical modeling.) The strong inferences we make about relationships between visual elements from relatively sparse visual information are called "gestalt rules." They are not pure perceptual effects like the checkerboard illusions. Rather, they describe our tendency to infer relationships between the objects we are looking at in a way that goes beyond what is strictly visible. Figure 1.21 provides some examples.

What sorts of relationships are inferred, and under what circumstances? In general we want to identify groupings, classifications, or entities than can be treated as the same thing or part of the same thing:

- *Proximity*: Things that are spatially near to one another seem to be related.
- *Similarity*: Things that look alike seem to be related.
- *Connection*: Things that are visually tied to one another seem to be related.
- *Continuity*: Partially hidden objects are completed into familiar shapes.
- *Closure*: Incomplete shapes are perceived as complete.
- *Figure and ground*: Visual elements are taken to be either in the foreground or in the background.
- *Common fate*: Elements sharing a direction of movement are perceived as a unit.

Some kinds of visual cues outweigh others. For example, in the upper left of figure 1.21, the circles are aligned horizontally into rows, but their proximity by column takes priority, and we see three groups of circles. In the upper right, the three groups are still salient but the row of blue circles is now seen as a grouped entity. In the middle row of the figure, the left side shows mixed grouping by shape, size, and color. Meanwhile the right side of the row shows that direct connection outweighs shape. Finally the two schematic plots in the bottom row illustrate both connection and common fate, in that the lines joining the shapes tend to be read left-to-right as part of a series. Note also the points in the lower right plot where the lines cross. There are gaps in the line segments joining the circles, but we perceive this as them "passing underneath" the lines joining the triangles.

1.4 Visual Tasks and Decoding Graphs

The workings of our visual system and our tendency to make inferences about relationships between visible elements form the basis of our ability to interpret graphs of data. There is more involved besides that, however. Beyond core matters of perception lies the question of interpreting and understanding particular kinds of graphs. The proportion of people who can read and correctly interpret a scatterplot is lower than you might think. At the intersection of perception and interpretation there are specific visual tasks that people need to perform in order to properly see the graph in front of them. To understand a scatterplot, for example, the viewer needs to know a lot of general information, such as what a variable is, what the x-y coordinate plane looks like, why we might want to compare two variables on it, and the convention of putting the supposed cause or "independent" variable on the x-axis. Even if viewers understand all these things, they must still perform the visual task of interpreting the graph. A scatterplot is a visual *representation* of data, not a way to magically transmit pure understanding. Even well-informed viewers may do worse than we think when connecting the picture to the underlying data (Doherty, et al. 2007; Rensink & Baldridge 2010).

In the 1980s William S. Cleveland and Robert McGill conducted some experiments identifying and ranking theses tasks for different types of graphics (Cleveland & McGill, 1984, 1987). Most

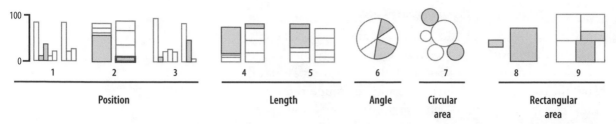

Figure 1.22: Schematic representation of basic perceptual tasks for nine chart types, by Heer and Bostock, following Cleveland and McGill. In both studies, participants were asked to make comparisons of highlighted portions of each chart type and say which was smaller.

often, research subjects were asked to estimate two values within a chart (e.g., two bars in a bar chart, or two slices of a pie chart) or compare values between charts (e.g., two areas in adjacent stacked bar charts). Cleveland went on to apply the results of this work, developing the trellis display system for data visualization in S, the statistical programming language developed at Bell Labs. (R is a later implementation of S.) He also wrote two excellent books that describe and apply these principles (Cleveland 1993, 1994).

In 2010 Heer & Bostock replicated Cleveland's earlier experiments and added a few assessments, including evaluations of rectangular-area graphs, which have become more popular in recent years. These include treemaps, where a square or rectangle is subdivided into further rectangular areas representing some proportion or percentage of the total. It looks a little like a stacked bar chart with more than one column. The comparisons and graph types made by their research subjects are shown schematically in figure 1.22. For each graph type, subjects were asked to identify the smaller of two marked segments on the chart and then to "make a quick visual judgment" estimating what percentage the smaller one was of the larger. As can be seen from the figure, the charts tested encoded data in different ways. Types 1–3 use position encoding along a common scale while types 4 and 5 use length encoding. The pie chart encodes values as angles, and the remaining charts as areas, using either circular, separate rectangles (as in a cartogram) or subrectangles (as in a treemap).

Their results are shown in figure 1.23, along with Cleveland and McGill's original results for comparison. The replication was quite good. The overall pattern of results seems clear, with performance worsening substantially as we move away from comparison on a common scale to length-based comparisons to angles and finally areas. Area comparisons perform even worse than the (justifiably) maligned pie chart.

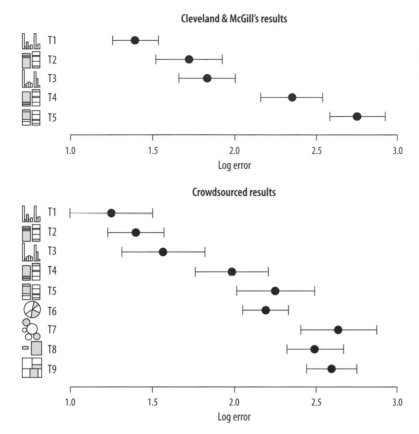

Figure 1.23: Cleveland and McGill's original results (*top*) and Heer and Bostock's replication with additions (*bottom*) for nine chart types.

These findings, and other work in this tradition, strongly suggest that there are better and worse ways of visually representing data when the task the user must perform involves estimating and comparing values within the graph. Think of this as a "decoding" operation that the viewer must perform in order to understand the content. The data values were encoded or mapped in to the graph, and now we have to get them back out again. When doing this, we do best judging the *relative position* of elements aligned on a *common scale*, as, for example, when we compare the heights of bars on a bar chart, or the position of dots with reference to a fixed x- or y-axis. When elements are not aligned but still share a scale, comparison is a little harder but still pretty good. It is more difficult again to compare the lengths of lines without a common baseline.

Outside of position and length encodings, things generally become harder and the decoding process is more error prone. We tend to misjudge quantities encoded as *angles*. The size of acute angles tends to be underestimated, and the size of obtuse angles overestimated. This is one reason pie charts are usually a bad idea. We also misjudge *areas* poorly. We have known for a long time that area-based comparisons of quantities are easily misinterpreted or exaggerated. For example, values in the data might be encoded as lengths, which are then squared to make the shape on the graph. The result is that the difference in size between the squares or rectangles area will be much larger than the difference between the two numbers they represent.

Comparing the areas of circles is prone to more error again, for the same reason. It is possible to offset these problems somewhat by choosing a more sophisticated method for encoding the data as an area. Instead of letting the data value be the length of the side of a square or the radius of the circle, for example, we could map the value directly to area and back-calculate the side length or radius. Still, the result will generally not be as good as alternatives. These problems are further compounded for "three-dimensional" shapes like blocks, cylinders, or spheres, which appear to represent volumes. And as saw with the 3-D bar chart in figure 1.10, the perspective or implied viewing angle that accompanies these kinds of charts creates other problems when it comes to reading the scale on a y-axis.

Finally, we find it hard to judge *changes in slope*. The estimation of rates of change in lines or trends is strongly conditioned by the aspect ratio of the graph, as we saw in figure 1.12. Our relatively weak judgment of slopes also interacts badly with three-dimensional representations of data. Our ability to scan the "away" dimension of depth (along the z-axis) is weaker than our ability to scan the x- and y-axes. For this reason, it can be disproportionately difficult to interpret data displays of point clouds or surfaces displayed with three axes. They can look impressive, but they are also harder to grasp.

1.5 Channels for Representing Data

Graphical elements represent our data in ways that we can see. Different sorts of variables attributes can be represented more or less well by different kinds of visual marks or representations, such as points, lines, shapes, and colors. Our task is to come up with methods that encode or map variables in the right way. As we do this, we face several constraints. First, the channel or mapping that we choose needs to be capable of representing the kind of data that we have. If we want to pick out unordered categories, for example, choosing a continuous gradient to represent them will not make much sense. If our variable is continuous, it will not be helpful to represent it as a series of shapes.

Second, given that the data *can* be comprehensibly represented by the visual element we choose, we will want to know how effective that representation is. This was the goal of Cleveland's research. Following Tamara Munzer (2014, 101–3), Figures 1.24 and 1.25 present an approximate ranking of the effectiveness of different channels for ordered and unordered data, respectively. If we have ordered data and we want the viewer to efficiently make comparisons, then we should try to encode it as a position on a common scale. Encoding numbers as lengths (absent a scale) works too, but not as effectively. Encoding them as areas will make comparisons less accurate again, and so on.

Third, the effectiveness of our graphics will depend not just on the channel that we choose but on the perceptual details of how we implement it. So, if we have a measure with four categories ordered from lowest to highest, we might correctly decide to represent it using a sequence of colors. But if we pick the wrong sequence, the data will still be hard to interpret, or actively misleading. In a similar way, if we pick a bad set of hues for an unordered categorical variable, the result might not just be unpleasant to look at but actively misleading.

Finally, bear in mind that these different channels or mappings for data are not in themselves kinds of graphs. They are just the elements or building blocks for graphs. When we choose to encode a variable as a position, a length, an area, a shade of gray, or a color, we have made an important decision that narrows down what the resulting plot can look like. But this is not the same as deciding what type of plot it will be, in the sense of choosing whether to make a dotplot or a bar chart, a histogram or a frequency polygon, and so on.

1.6 Problems of Honesty and Good Judgment

Figure 1.26 shows two ways of redrawing our life expectancy figure (fig. 1.4). Each of these plots is far less noisy than the junk-filled monstrosity we began with. But they also have design features that

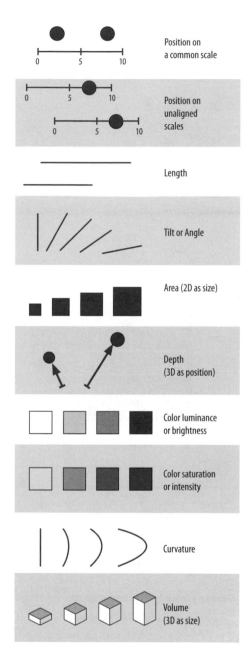

Figure 1.24: Channels for mapping ordered data (continuous or other quantitative measures), arranged top to bottom from more to less effective, after Munzer (2014, 102).

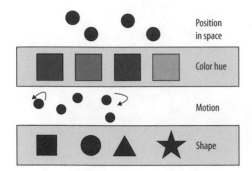

Figure 1.25: Channels for mapping unordered categorical data, arranged top-to-bottom from more to less effective, after Munzer (2014, 102).

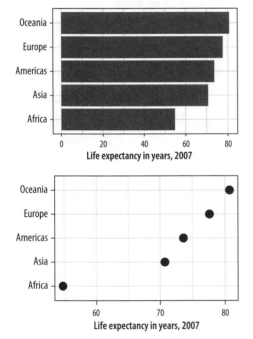

Figure 1.26: Two simpler versions of our junk chart. The scale on the bar chart version goes to zero, while the scale on the dotplot version is confined to the range of values taken by the observations.

could be argued over and might even matter substantively depending on the circumstances. For example, consider the scales on the x-axis in each case. The left-hand panel in figure 1.26 is a bar chart, and the length of the bar represents the value of the variable "average life expectancy in 2007" for each continent. The scale starts at zero and extends to just beyond the level of the largest value. Meanwhile the right-hand panel is a Cleveland dotplot. Each observation is represented by a point, and the scale is restricted to the range of the data as shown.

It is tempting to lay down inflexible rules about what to do in terms of producing your graphs, and to dismiss people who don't follow them as producing junk charts or lying with statistics. But being honest with your data is a bigger problem than can be solved by rules of thumb about making graphs. In this case there is a moderate level of agreement that bar charts should generally include a zero baseline (or equivalent) given that bars make lengths salient to the viewer. But it would be a mistake to think that a dotplot was by the same token deliberately misleading, just because it kept itself to the range of the data instead.

Which one is to be preferred? It is tricky to give an unequivocal answer, because the reasons for preferring one type of scaling over another depend in part on how often people actively try to mislead others by preferring one sort of representation over another. On the one hand, there is a lot of be said in favor of showing the data over the range we observe it, rather than forcing every scale to encompass its lowest and highest theoretical value. Many otherwise informative visualizations would become useless if it was mandatory to include a zero point on the x- or y-axis. On the other hand, it's also true that people sometimes go out of their way to restrict the scales they display in a way that makes their argument look better. Sometimes this is done out of active malice, other times out of passive bias, or even just a hopeful desire to see what you want to see in the data. (Remember, often the main audience for your visualizations is yourself.) In those cases, the resulting graphic will indeed be misleading.

Rushed, garish, and deliberately inflammatory or misleading graphics are a staple of social media sharing and the cable news cycle. But the problem comes up in everyday practice as well, and the two can intersect if your work ends up in front of a public audience. For example, let's take a look at some historical data on law school enrollments. A decline in enrollments led to some reporting on trends since the early 1970s. The results are shown in figure 1.27.

The first panel shows the trend in the number of students beginning law school each year since 1973. The y-axis starts from just below the lowest value in the series. The second panel shows the same data but with the y-axis minimum set to zero instead. The columnist and writer Justin Fox saw the first version and remarked on how amazing it was. He was then quite surprised at the strong reactions he got from people who insisted the y-axis should have included zero. The original chart was "possibly … one of the worst represented charts I've ever seen," said one interlocutor. Another remarked that "graphs that don't go to zero are a thought crime" (Fox 2014).

My own view is that the chart without the zero baseline shows you that, after almost forty years of mostly rising enrollments, law school enrollments dropped suddenly and precipitously around 2011 to levels not seen since the early 1970s. The levels are clearly labeled, and the decline does look substantively surprising and significant. In a well-constructed chart the axis labels are a necessary guide to the reader, and we should expect readers to pay attention to them. The chart with the zero baseline, meanwhile, does not add much additional information beyond reminding you, at the cost of wasting some space, that 35,000 is a number quite a lot larger than zero.

That said, I am sympathetic to people who got upset at the first chart. At a minimum, it shows they know to read the axis labels on a graph. That is less common than you might think. It likely also shows they know interfering with the axes is one way to make a chart misleading, and that it is not unusual for that sort of thing to be done deliberately.

1.7 Think Clearly about Graphs

I am going to assume that your goal is to draw effective graphs in an honest and reproducible way. Default settings and general rules of good practice have limited powers to stop you from doing the wrong thing. But one thing they can do is provide not just tools for making graphs but also a framework or set of concepts that helps you think more clearly about the good work you want to produce. When learning a graphing system or toolkit, people

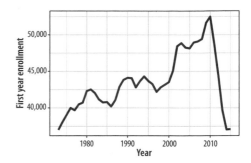

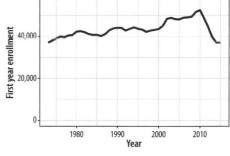

Figure 1.27: Two views of the rapid decline in law school enrollments in the mid-2010s.

often start thinking about specific ways they want their graph to look. They quickly start formulating requests. They want to know how to make a particular kind of chart, or how to change the typeface for the whole graph, or how to adjust the scales, or how to move the title, customize the labels, or change the colors of the points.

These requests involve different features of the graph. Some have to do with basic features of the figure's structure, with which bits of data are encoded as or mapped to elements such as shape, line, or color. Some have to do with the details of how those elements are represented. If a variable is mapped to shape, which shapes will be chosen, exactly? If another variable is represented by color, which colors in particular will be used? Some have to do with the framing or guiding features of the graph. If there are tickmarks on the x-axis, can I decide where they should be drawn? If the chart has a legend, will it appear to the right of the graph or on top? If data points have information encoded in both shape and color, do we need a separate legend for each encoding, or can we combine them into a single unified legend? And some have to do with thematic features of the graph that may greatly affect how the final result looks but are not logically connected to the structure of the data being represented. Can I change the title font from Times New Roman to Helvetica? Can I have a light blue background in all my graphs?

A real strength of ggplot is that it implements a *grammar of graphics* to organize and make sense of these different elements (Wilkinson 2005). Instead of a huge, conceptually flat list of options for setting every aspect of a plot's appearance at once, ggplot breaks up the task of making a graph into a series of distinct tasks, each bearing a well-defined relationship to the structure of the plot. When you write your code, you carry out each task using a function that controls that part of the job. At the beginning, ggplot will do most of the work for you. Only two steps are required. First, you must give some information to the `ggplot()` function. This establishes the core of the plot by saying what data you are using and what variables will be linked or *mapped* to features of the plot. Second, you must choose a `geom_` function. This decides what sort of plot will be drawn, such as a scatterplot, a bar chart, or a boxplot.

As you progress, you will gradually use other functions to gain more fine-grained control over other features of the plot, such as scales, legends, and thematic elements. This also means that, as you learn ggplot, it is very important to grasp the core steps first, before worrying about adjustments and polishing. And so that is how we'll proceed. In the next chapter we will learn how to get up and running in R and make our first graphs. From there, we will work through examples that introduce each element of ggplot's way of doing things. We will be producing sophisticated plots quite quickly, and we will keep working on them until we are in full control of what we are doing. As we go, we will learn about some ideas and associated techniques and tricks to make R do what we want.

1.8 Where to Go Next

For an entertaining and informative overview of various visual effects and optical "illusions," take a look at Michael Bach's website at michaelbach.de. If you would like to learn more about the relationship between perception and data visualization, follow up on some of the references in this chapter. Munzer (2014), Ware (2008), and Few (2009) are good places to start. William Cleveland's books (1993, 1994) are models of clarity and good advice. As we shall see beginning in the next chapter, the ideas developed in Wilkinson (2005) are at the heart of ggplot's approach to visualization. Finally, foundational work by Bertin (2010) lies behind a lot of thinking on the relationship between data and visual elements.

2 Get Started

In this chapter, we will begin to learn how to create pictures of data that people, including ourselves, can look at and learn from. R and ggplot are the tools we will use. The best way to learn them is to follow along and repeatedly write code as you go. The material in this book is designed to be interactive and hands-on. If you work through it with me using the approach described below, you will end up with a book much like this one, with many code samples alongside your notes, and the figures or other output produced by that code shown nearby.

I strongly encourage you to type out your code rather than copying and pasting the examples from the text. Typing it out will help you learn it. At the beginning it may feel like tedious transcription you don't fully understand. But it slows you down in a way that gets you used to what the syntax and structure of R are like and is a very effective way to learn the language. It's especially useful for ggplot, where the code for our figures will repeatedly have a similar structure, built up piece by piece.

2.1 Work in Plain Text, Using RMarkdown

When taking notes, and when writing your own code, you should write plain text in a text editor. Do not use Microsoft Word or some other word processor. You may be used to thinking of your final outputs (e.g., a Word file, a PDF document, presentation slides, or the tables and figures you make) as what's "real" about your project. Instead, it's better to think of the data and code as what's real, together with the text you write. The idea is that all your finished output—your figures, tables, text, and so on—can be procedurally and reproducibly generated from code, data, and written material stored in a simple, plain-text format.

The ability to reproduce your work in this way is important to the scientific process. But you should also see it as a pragmatic choice that will make life easier for you in the future. The reality for most of us is that the person who will most want to easily reproduce your work is *you*, six months or a year from now. This is especially true for graphics and figures. These often have a "finished" quality to them, as a result of much tweaking and adjustments to the details of the figure. That can make them hard to reproduce later. While it is normal for graphics to undergo a substantial amount of polishing on their way to publication, our goal is to do as much of this as possible *programmatically*, in code we write, rather than in a way that is retrospectively invisible, as, for example, when we edit an image in an application like Adobe Illustrator.

While learning ggplot, and later while doing data analysis, you will find yourself constantly pinging back and forth between three things:

1. *Writing code*. You will write a lot of code to produce plots. You will also write code to load your data and to look quickly at tables of that data. Sometimes you will want to summarize, rearrange, subset, or augment your data, or run a statistical model with it. You will want to be able to write that code as easily and effectively as possible.
2. *Looking at output*. Your code is a set of instructions that, when executed, produces the output you want: a table, a model, or a figure. It is often helpful to be able to see that output and its partial results. While we're working, it's also useful to keep the code and the things produced by the code close together, if we can.
3. *Taking notes*. You will also be writing about what we are doing and what your results mean. When learning how to do something in ggplot, for instance, you will want to make notes to yourself about what you did, why you wrote it this way rather than that, or what this new concept, function, or instruction does. Later, when doing data analysis and making figures, you will be writing up reports or drafting papers.

How can you do all this effectively? The simplest way to keep code and notes together is to write your code and intersperse it with comments. All programming languages have some way of demarcating lines as comments, usually by putting a special character (like #) at the start of the line. We could create a plain-text script file called, e.g., notes.r, containing code and our comments on it. This is fine as far as it goes. But except for very short files, it

Markdown	Output
# Header	**Header**
## Subhead	**Subhead**
Plain text	Plain text
italics	*italics*
bold	**bold**
`verbatim`	verbatim
1. List	1. List
2. List	2. List
- Bullet 1	° Bullet 1
- Bullet 2	° Bullet 2
Footnote.[^1]	Footnote[1]
[^1]: The footnote.	[1]The footnote.

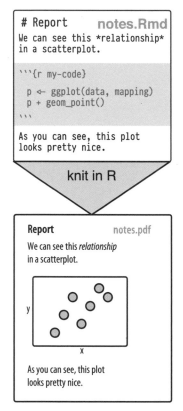

Figure 2.1: *Top:* Some elements of RMarkdown syntax. *Bottom:* From a plain text RMarkdown file to PDF output.

The format is language agnostic and can be used with, e.g., Python and other languages.

will be difficult to do anything useful with the comments we write. If we want a report from an analysis, for example, we will have to write it up separately. While a script file can keep comments and code together, it loses the connection between code and its output, such as the figure we want to produce. But there is a better alternative: we can write our notes using RMarkdown.

An RMarkdown file is just a plain text document where text (such as notes or discussion) is interspersed with pieces, or *chunks*, of R code. When you feed the document to R, it *knits* this file into a new document by running the R code piece by piece, in sequence, and either supplementing or replacing the chunks of code with their output. The resulting file is then converted into a more readable document formatted in HTML, PDF, or Word. The noncode segments of the document are plain text, but they can have simple formatting instructions in them. These are set using Markdown, a set of conventions for marking up plain text in a way that indicates how it should be formatted. The basic elements of Markdown are shown in the upper part of figure 2.1. When you create a markdown document in R Studio, it contains some sample text to get you started.

RMarkdown documents look like the one shown schematically in the lower part of figure 2.1. Your notes or text, with Markdown formatting as needed, are interspersed with code. There is a set format for code chunks. They look like this:

```
```{r}


```
```

Three backticks (on a U.S. keyboard, that's the character under the escape key) are followed by a pair of curly braces containing the name of the language we are using. The backticks-and-braces part signals that a chunk of code is about to begin. You write your code as needed and then end the chunk with a new line containing just three backticks.

If you keep your notes in this way, you will be able to see the code you wrote, the output it produces, and your own commentary or clarification on it in a convenient way. Moreover, you can turn it into a good-looking document right away.

```
R version 3.4.1 (2017-06-30) -- "Single Candle"
Copyright (C) 2017 The R Foundation for Statistical Computing
Platform: x86_64-apple-darwin15.6.0 (64-bit)

R is free software and comes with ABSOLUTELY NO WARRANTY.
You are welcome to redistribute it under certain conditions.
Type 'license()' or 'licence()' for distribution details.

  Natural language support but running in an English locale

R is a collaborative project with many contributors.
Type 'contributors()' for more information and
'citation()' on how to cite R or R packages in publications.

Type 'demo()' for some demos, 'help()' for on-line help, or
'help.start()' for an HTML browser interface to help.
Type 'q()' to quit R.

> ▌
```

Figure 2.2: Bare-bones R running from the Terminal.

2.2 Use R with RStudio

The RStudio environment

R itself is a relatively small application with next to no user interface. Everything works through a command line, or *console*. At its most basic, you launch it from your Terminal application (on a Mac) or Command Prompt (on Windows) by typing R. Once launched, R awaits your instructions at a command line of its own, denoted by the right angle bracket symbol, > (fig. 2.2). When you type an instruction and hit return, R interprets it and sends any resulting output back to the console.

In addition to interacting with the console, you can also write your code in a text file and send that to R all at once. You can use any good text editor to write your .r scripts. But although a plain text file and a command line are the absolute minimum you need to work with R, it is a rather spartan arrangement. We can make life easier for ourselves by using RStudio. RStudio is an "integrated development environment," or IDE. It is a separate application from R proper. When launched, it starts up an instance of R's console inside of itself. It also conveniently pulls together various other elements to help you get your work done. These include the document where you are writing your code, the output it produces, and R's help system. RStudio also knows about RMarkdown and understands a lot about the R language and the organization of your project. When you launch RStudio, it should look much like figure 2.3.

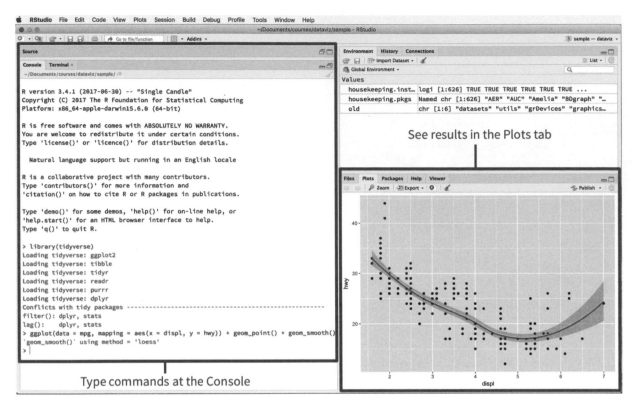

Figure 2.3: The RStudio IDE.

Create a project

You can create your new project wherever you like—most commonly it will go somewhere in your Documents folder.

To begin, create a project. From the menu, choose File > New Project … from the menu bar, choose the New Directory option, and create the project. Once it is set up, create an RMarkdown file in the directory, with File > New File > RMarkdown. This will give you a set of choices including the default "Document." The socviz library comes with a small RMarkdown template that follows the structure of this book. To use it instead of the default document, after selecting File > New File > RMarkdown, choose the "From Template" option in the sidebar of the dialog box that appears. Then choose "Data Visualization Notes" from the resulting list of options. When the RMarkdown document appears, save it right away in your project folder, with File > Save. The socviz template contains a information about how RMarkdown works, together with some headers to get you started. Read what it has to say. Look at the code chunks and RMarkdown formatting. Experiment with knitting the document, and compare

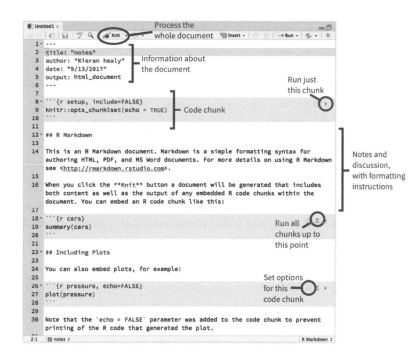

Figure 2.4: An RMarkdown file open in R Studio. The small icons in the top right-hand corner of each code chunk can be used to set options (the gear icon), run all chunks up to the current one (the downward-facing triangle), and just run the current chunk (the right-facing triangle).

the output to the content of the plain text document. Figure 2.4 shows you what an RMarkdown file looks like when opened in RStudio.

RMarkdown is not required for R. An alternative is to use an R script, which contains R commands only. R script files conventionally have the extension .r or .R. (RMarkdown files conventionally end in .Rmd.) A brief project might just need a single .r file. But RMarkdown is useful for documents, notes, or reports of any length, especially when you need to take notes. If you do use an .r file you can leave comments or notes to yourself by starting a line with the hash character, #. You can also add comments at the end of lines in this way, as for any particular line R will ignore whatever code or text that appears after a #.

RStudio has various keyboard and menu shortcuts to help you edit code and text quickly. For example, you can insert chunks of code in your RMarkdown document with a keyboard shortcut. This saves you from writing the backticks and braces every time. You can run the current line of code with a shortcut, too. A third shortcut gives you a pop-over display with summary of many other useful keystroke combinations. RMarkdown documents can include all kinds of other options and formatting paraphernalia,

You can create an r script via File > New File > R Script.

Command+Option+I on MacOS. Ctrl+Alt+I on Windows.
Command+Enter on MacOS. Alt+Enter on Windows.

Option+Shift+K on MacOS. Alt-Shift-K on Windows.

from text formatting to cross-references to bibliographical information. But never mind about those for now.

To make sure you are ready to go, load the `tidyverse`. The tidyverse is a suite of related packages for R developed by Hadley Wickham and others. The `ggplot2` package is one of its components. The other pieces make it easier to get data into R and manipulate it once it is there. Either knit the notes file you created from the socviz template or load the packages manually at the console:

```
library(tidyverse)

library(socviz)
```

Load the `socviz` package after the tidyverse. This library contains datasets that we will use throughout the book, along with some other tools that will make life easier. If you get an error message saying either package can't be found, then reread the "Before You Begin" section in the preface to this book and follow the instructions there.

You need to *install* a package only once, but you will need to *load* it at the beginning of each R session with `library()` if you want to use the tools it contains. In practice this means that the very first lines of your working file should contain a code chunk that loads the packages you will need in the file. If you forget to do this, then R will be unable to find the functions you want to use later on.

2.3 Things to Know about R

Any new piece of software takes a bit of getting used to. This is especially true when using an IDE in a language like R. You are getting oriented to the language itself (what happens at the console) while learning to take notes in what might seem like an odd format (chunks of code interspersed with plain-text comments), in an IDE that has many features designed to make your life easier in the long run but which can be hard to decipher at the beginning. Here are some general points to bear in mind about how R is designed. They might help you get a feel for how the language works.

Everything has a name

In R, everything you deal with has a name. You refer to things by their names as you examine, use, or modify them. Named entities include variables (like x or y), data that you have loaded (like my_data), and functions that you use. (More about functions momentarily.) You will spend a lot of time talking about, creating, referring to, and modifying things with names.

Some names are forbidden. These include reserved words like FALSE and TRUE, core programming words like Inf, for, else, break, and function, and words for special entities like NA and NaN. (These last two are codes designating missing data and "Not a Number," respectively.) You probably won't use these names by accident, but it's good do know that they are not allowed.

Some names you should not use even if they are technically permitted. These are mostly words that are already in use for objects or functions that form part of the core of R. These include the names of basic functions like q() or c(), common statistical functions like mean(), range(), or var(), and built-in mathematical constants like pi.

Names in R are case sensitive. The object my_data is not the same as the object My_Data. When choosing names for things, be concise, consistent, and informative. Follow the style of the tidyverse and name things in lowercase, separating words with the underscore character, _, as needed. Do not use spaces when naming things, including variables in your data.

Everything is an object

Some objects are built in to R, some are added via packages, and some are created by the user. But almost everything is some kind of object. The code you write will create, manipulate, and use named objects as a matter of course. We can start immediately. Let's create a vector of numbers. The command c() is a function. It's short for "combine" or "concatenate." It will take a sequence of comma-separated things inside the parentheses and join them together into a vector where each element is still individually accessible.

```
c(1, 2, 3, 1, 3, 5, 25)
```

```
## [1]  1  2  3  1  3  5 25
```

You can type the arrow using < and then -.

Instead of sending the result to the console, we can instead *assign* it to an object we create:

```
my_numbers ← c(1, 2, 3, 1, 3, 5, 25)
your_numbers ← c(5, 31, 71, 1, 3, 21, 6)
```

To see what you made, type the name of the object and hit return:

```
my_numbers
```

```
## [1]  1  2  3  1  3  5 25
```

Each of our numbers is still there and can be accessed directly if we want. They are now just part of a new object, a vector, called my_numbers.

You create objects by assigning them to names. The *assignment operator* is ←. Think of assignment as the verb "gets," reading left to right. So the bit of code above can be read as "The object my_numbers gets the result of concatenating the following numbers: 1, 2, ..." The operator is two separate keys on your keyboard: the < key and the – (minus) key. Because you type this so often in R, there is a shortcut for it in R Studio. To write the assignment operator in one step, hold down the option key and hit –. On Windows hold down the alt key and hit –. You will be constantly creating objects in this way, and trying to type the two characters separately is both tedious and prone to error. You will make hard-to-notice mistakes like typing < – (with a space in between the characters) instead of ←.

If you learn only one keyboard shortcut in RStudio, make it this one! Always use Option+minus on MacOS or Alt+minus on Windows to type the assignment operator.

When you create objects by assigning things to names, they come into existence in R's *workspace* or *environment*. You can think of this most straightforwardly as your project directory. Your workspace is specific to your current project. It is the folder from which you launched R. Unless you have particular needs (such as extremely large datasets or analytical tastes that take a very long time) you will not need to give any thought to where objects "really" live. Just think of your code and data files as the permanent features of your project. When you start up an R project, you will generally begin by loading your data. That is, you will read it in from disk and assign it to a named object like my_data. The rest of your code will be a series of instructions to act on and create more named objects.

You do things using functions

You do almost everything in R using functions. Think of a function as a special kind of object that can perform actions for you. It produces output based on the input it receives. When we want a function to do something for us, we *call* it. It will reliably do what we tell it. We give the function some information, it acts on that information, and some results come out the other side. A schematic example is shown in fig. 2.5. Functions can be recognized by the parentheses at the end of their names. This distinguishes them from other objects, such as single numbers, named vectors, and tables of data.

The parentheses are what allow you to send information to the function. Most functions accept one or more named *arguments*. A function's arguments are the things it needs to know in order to do something. They can be some bit of your data (data = my_numbers), or specific instructions (title = "GDP per Capita"), or an option you want to choose (smoothing = "splines", show = FALSE). For example, the object my_numbers is a numeric vector:

```
my_numbers
```

```
## [1]  1  2  3  1  3  5 25
```

But the thing we used to create it, c(), is a function. It concatenates items into a vector composed of the series of comma-separated elements you give it. Similarly, mean() is a function that calculates a simple average for a vector of numbers. What happens if we just type mean() without any arguments inside the parentheses?

```
mean()
# Error in mean.default() : argument 'x' is
# missing, with no default
```

The error message is terse but informative. The function needs an argument to work, and we haven't given it one. In this case, 'x', the name of another object that mean() can perform its calculation on:

```
fn_name( argument1 = <value1>,
         argument2 = <value2>,
         argument3 = <value3>)


plot_it( xvals = my_numbers,
         yvals = your_numbers,
         title = "Our Number Plot")
```

Figure 2.5: *Upper:* What functions look like, schematically. *Lower:* an imaginary function that takes two vectors and plots them with a title. We supply the function with the particular vectors we want it to use, and the title. The vectors are object so are given as is. The title is not an object so we enclose it in quotes.

```
mean(x = my_numbers)
```

```
## [1] 5.714286
```

```
mean(x = your_numbers)
```

```
## [1] 19.71429
```

While the function arguments have names that are used internally, (here, x), you don't strictly need to specify the name for the function to work:

```
mean(my_numbers)
```

```
## [1] 5.714286
```

See the appendix for a guide to how to read the help page for a function.

If you omit the name of the argument, R will just assume you are giving the function what it needs, and in a default order. The documentation for a function will tell you what the order of required arguments is for any particular function. For simple functions that require only one or two arguments, omitting their names is usually not confusing. For more complex functions, you will typically want to use the names of the arguments rather than try to remember what the ordering is.

In general, when providing arguments to a function the syntax is <argument> = <value>. If <value> is a named *object* that already exists in your workspace, like a vector of numbers or a table of data, then you provide it unquoted, as in mean(my_numbers). If <value> is not an object, a number, or a logical value like TRUE, then you usually put it in quotes, e.g., labels(x = "X Axis Label").

Functions take inputs via their arguments, do something, and return outputs. What the output is depends on what the function does. The c() function takes a sequence of comma-separated elements and returns a vector consisting of those same elements. The mean() function takes a vector of numbers and returns a single number, their average. Functions can return far more than single numbers. The output returned by functions can be a table of data, or a complex object such as the results of a linear model, or the instructions needed to draw a plot on the screen (as we shall see).

For example, the summary() function performs a series of calculations on a vector and produces what is in effect a little table with named elements.

A function's argument names are internal to that function. Say you have created an object in your environment named x, for example. A function like mean() also has a named argument, x, but R will not get confused by this. It will not use your x object by mistake.

As we have already seen with c() and mean(), you can assign the result of a function to an object:

```
my_summary ← summary(my_numbers)
```

When you do this, there's no output to the console. R just puts the results into the new object, as you instructed. To look inside the object you can type its name and hit return:

```
my_summary
```

```
##    Min. 1st. Qu.  Median   Mean 3rd Qu.    Max.
##    1.00    1.50    3.00   5.71    4.00   25.00
```

Functions come in packages

The code you write will be more or less complex depending on the task you want to accomplish. Once you have gotten used to working in R, you will probably end up writing your own functions to produce the results that you need. But as with other programming languages, you will not have to do everything yourself. Families of useful functions are bundled into packages that you can install, load into your R session, and make use of as you work. Packages save you from reinventing the wheel. They make it so that you do not, for example, have to figure out how to write code from scratch to draw a shape on screen, or load a data file into memory. Packages are also what allow you to build on the efforts of others in order to do your own work. Ggplot is a library of functions. There are many other such packages, and we will make use of several throughout this book, by either loading them with the library() function or "reaching in" to them and pulling a useful function from them directly. Writing code and functions of your own is a good way to get a sense of the amazing volume of effort put into R and its associated toolkits, work freely contributed by many hands over the years and available for anyone to use.

All the visualization we will do involves choosing the right function or functions and then giving those functions the right instructions through a series of named arguments. Most of the mistakes we will make, and the errors we will fix, will involve us having not picked the right function, or having not fed the function the right arguments, or having failed to provide information in a form the function can understand.

For now, just remember that you do things in R by creating and manipulating named objects. You manipulate objects by feeding information about them to functions. The functions do something useful with that information (calculate a mean, recode a variable, fit a model) and give you the results back.

```
table(my_numbers)
```

```
## my_numbers
##  1  2  3  5 25
##  2  1  2  1  1
```

```
sd(my_numbers)
```

```
## [1] 8.6
```

```
my_numbers * 5
```

```
## [1]   5  10  15   5  15  25 125
```

```
my_numbers + 1
```

```
## [1]  2  3  4  2  4  6 26
```

```
my_numbers + my_numbers
```

```
## [1]  2  4  6  2  6 10 50
```

The first two functions here gave us a simple table of counts and calculated the standard deviation of my_numbers. It's worth noticing what R did in the last three cases. First we multiplied my_numbers by five. R interprets that as you asking it to take each *element* of my_numbers one at a time and multiply it by five. It does the same with the instruction my_numbers + 1. The single value is "recycled" down the length of the vector. By contrast, in the last case we add `my_numbers` to itself. Because the two objects being added are the same length, R adds each element in the first vector to the corresponding element in the second vector. This is an example of a *vectorized* operation.

If you're not sure what an object is, ask for its class

Every object has a *class*. This is the sort of object it is, whether a vector, a character string, a function, a list, and so on. Knowing an object's class tells you a lot about what you can and can't do with it.

```
class(my_numbers)
```

```
## [1] "numeric"
```

```
class(my_summary)
```

```
## [1] "summaryDefault" "table"
```

```
class(summary)
```

```
## [1] "function"
```

Certain actions you take may change an object's class. For instance, consider `my_numbers` again:

```
my_new_vector <- c(my_numbers, "Apple")
my_new_vector
```

```
## [1] "1"    "2"    "3"    "1"    "3"    "5"    "25"
## [8] "Apple"
```

```
class(my_new_vector)
```

```
## [1] "character"
```

The function added the word "Apple" to our vector of numbers, as we asked. But in doing so, the result is that the new object also has a new class, switching from "numeric" to "character." All the numbers are now enclosed in quotes. They have been turned into character strings. In that form, they can't be used in calculations.

Most of the work we'll be doing will not involve directly picking out this or that value from vectors or other entities. Instead we will try to work at a slightly higher level that will be easier and safer. But it's worth knowing the basics of how elements of vectors can be referred to because the c() function in particular is a useful tool.

We will spend a lot of time in this book working with a series of datasets. These typically start life as files stored locally on your computer or somewhere remotely accessible to you. Once they are imported into R, then like everything else they exist as objects of some kind. R has several classes of objects used to store data. A basic one is a matrix, which consists of rows and columns of numbers. But the most common kind of data object in R is a *data frame*, which you can think of as a rectangular table consisting of rows (of observations) and columns (of variables). In a data frame the columns can be of different classes. Some may be character strings, some numeric, and so on. For instance, here is a very small dataset from the socviz library:

The appendix has a little more discussion of the basics of selecting the elements within objects.

```
titanic
```

```
##       fate    sex    n percent
## 1 perished   male 1364    62.0
## 2 perished female  126     5.7
## 3 survived   male  367    16.7
## 4 survived female  344    15.6
```

```
class(titanic)
```

```
## [1] "data.frame"
```

In this titanic data, two of the columns are numeric and two are not. You can access the rows and columns in various ways. For example, the $ operator allows you to pick out a named column of a data frame:

```
titanic$percent
```

```
## [1] 62.0  5.7 16.7 15.6
```

Appendix 1 contains more information about selecting particular elements from different kinds of objects.

We will also regularly encounter a slightly augmented version of a data frame called a *tibble*. The tidyverse libraries make extensive use of tibbles. Like data frames, they are used to store variables of different classes all together in a single table of data. They also do a little more to let us know about what they contain and are a little more friendly when interacted with from the console. We can convert a data frame to a tibble if we want:

```
titanic_tb ← as_tibble(titanic)
titanic_tb
```

```
## # A tibble: 4 x 4
##   fate      sex         n percent
##   <fct>    <fct>   <dbl>  <dbl>
## 1 perished male   1364.   62.0
## 2 perished female  126.    5.70
## 3 survived male    367.   16.7
## 4 survived female  344.   15.6
```

Look carefully at the top and bottom of the output to see what additional information the tibble class gives you over and above the data frame version.

To see inside an object, ask for its structure

The str() function is sometimes useful. It lets you see what is inside an object.

```
str(my_numbers)
```

```
##  num [1:7] 1 2 3 1 3 5 25
```

```
str(my_summary)
```

```
## Classes 'summaryDefault', 'table'
 Named num [1:6] 1 1.5 3 5.71 4 ...
##   ..- attr(*, "names")= chr [1:6] "Min."
 "1st Qu." "Median" "Mean" ...
```

Fair warning: while some objects are relatively simple (a vector is just a sequence of numbers), others are more complicated, so asking about their str() might output a forbidding amount of information to the console. In general, complex objects are

RStudio also has the ability to summarize and look inside objects, via its Environment tab.

organized collections of simpler objects, often assembled as a big list, sometimes with a nested structure. Think, for example, of a master to-do list for a complex activity like moving house. It might be organized into subtasks of different kinds, several of which would themselves have lists of individual items. One list of tasks might be related to scheduling the moving truck, another might cover things to be donated, and a third might be related to setting up utilities at the new house. In a similar way, the objects we create to make plots will have many parts and subparts, as the overall task of drawing a plot has many individual to-do items. But we will be able to build these objects up from simple forms through a series of well-defined steps. And unlike moving house, the computer will take care of actually carrying out the task for us. We just need to get the to-do list right.

2.4 Be Patient with R, and with Yourself

Like all programming languages, R does exactly what you tell it to, rather than exactly what you want it to. This can make it frustrating to work with. It is as if one had an endlessly energetic, powerful, but also extremely literal-minded robot to order around. Remember that no one writes fluent, error-free code on the first go all the time. From simple typos to big misunderstandings, mistakes are a standard part of the activity of programming. This is why error checking, debugging, and testing are also a central part of programming. So just try to be patient with yourself and with R while you use it. Expect to make errors, and don't worry when that happens. You won't break anything. Each time you figure out why a bit of code has gone wrong, you will have learned a new thing about how the language works.

Here are three specific things to watch out for:

- Make sure parentheses are balanced and that every opening "(" has a corresponding closing ")".
- Make sure you complete your expressions. If you think you have completed typing your code but instead of seeing the > command prompt at the console you see the + character, that may mean R thinks you haven't written a complete expression yet. You can hit Esc or Ctrl-C to force your way back to the console and try typing your code again.

- In ggplot specifically, as you will see, we will build up plots a piece at a time by adding expressions to one another. When doing this, make sure your + character goes at the end of the line and not the beginning. That is, write this:

```
ggplot(data = mpg, aes(x = displ, y = hwy)) + geom_point()
```

and not this:

```
ggplot(data = mpg, aes(x = displ, y = hwy))
+geom_point()
```

R Studio will do its best to help you with the task of writing your code. It will highlight your code by syntax; it will try to match characters (like parentheses) that need to be balanced; it will try to narrow down the source of errors in code that fails to run; it will try to auto-complete the names of objects you type so that you make fewer typos; it will make help files more easily accessible and the arguments of functions directly available. Go slowly and see how the software is trying to help you out.

2.5 Get Data into R

Before we can plot anything at all, we have to get our data into R in a format it can use. Cleaning and reading in your data is one of the least immediately satisfying pieces of an analysis, whether you use R, Stata, SAS, SPSS, or any other statistical software. This is the reason that many of the datasets for this book are provided in a preprepared form via the socviz library rather than as data files you must manually read in. However, it is something you will have to face sooner rather than later if you want to use the skills you learn in this book. We might as well see how to do it now. Even when learning R, it can be useful and motivating to try out the code on your own data rather than working with the sample datasets.

Use the read_csv() function to read in comma-separated data. This function is in the readr package, one of the pieces of the tidyverse. R and the tidyverse also have functions to import various Stata, SAS, and SPSS formats directly. These can be found in the haven package. All we need to do is point read_csv() at a

These are all commercial software applications for statistical analysis. Stata, in particular, is in wide use across the social sciences.

file. This can be a local file, e.g., in a subdirectory called data/, or it can be a remote file. If read_csv() is given a URL or ftp address, it will follow it automatically. In this example, we have a CSV file called organdonation.csv stored at a trusted remote location. While online, we assign the URL for the file to an object, for convenience, and then tell read_csv() to fetch it for us and put it in an object named organs.

```
url ← "https://cdn.rawgit.com/kjhealy/viz-organdata/master/organdonation.csv"

organs ← read_csv(file = url)
```

```
## Parsed with column specification:
## cols(
##    .default = col_character(),
##    year = col_integer(),
##    donors = col_double(),
##    pop = col_integer(),
##    pop.dens = col_double(),
##    gdp = col_integer(),
##    gdp.lag = col_integer(),
##    health = col_double(),
##    health.lag = col_double(),
##    pubhealth = col_double(),
##    roads = col_double(),
##    cerebvas = col_integer(),
##    assault = col_integer(),
##    external = col_integer(),
##    txp.pop = col_double()
## )
```

```
## See spec(...) for full column specifications.
```

The resulting message at the console tells us the read_csv() function has assigned a class to each column of the object it created from the CSV file. There are columns with integer values, some are character strings, and so on. (The double class is for numbers other than integers.) Part of the reason read_csv() is telling you this information is that it is helpful to know what class each column, or variable, is. A variable's class determines what sort of operations can be performed on it. You also see this information because the tidyverse's read_csv() (with an underscore character in the middle of its name) is more opinionated than an older, and also still widely used, function, read.csv() (with a period in the middle of its name). The newer read_csv() will not classify variables as factors unless you tell it to. This is in contrast to the older function, which treats any vector of characters as a factor unless told otherwise. Factors have some very useful features in R (especially when it comes to representing various kinds of treatment and control groups in experiments), but they often trip up users who are not fully aware of them. Thus read_csv() avoids them unless you explicitly say otherwise.

R can read in data files in many different formats. The haven package provides functions to read files created in a variety of commercial software packages. If your dataset is a Stata .dta file, for instance, you can use the read_dta() function in much the same way as we used read_csv() above. This function can read and write variables stored as logical values, integers, numbers, characters, and factors. Stata also has a *labeled* data class that the haven library partially supports. In general you will end up converting labeled variables to one of R's basic classes. Stata also supports an extensive coding scheme for missing data. This is generally not used directly in R, where missing data is coded simply as NA. Again, you will need to take care that any labeled variables imported into R are coded properly, so that you do not end up mistakenly using missing data in your analysis.

When preparing your data for use in R, and in particular for graphing with ggplot, bear in mind that it is best if it is represented in a "tidy" format. Essentially this means that your data should be in *long* rather than *wide* format, with every observation a row and every variable a column. We will discuss this in more detail in chapter 3, and you can also consult the discussion of tidy data in the appendix.

R can also talk directly to databases, a topic not covered here.

See haven's documentation for more details.

2.6 Make Your First Figure

That's enough ground clearing for now. Writing code can be frustrating, but it also allows you to do interesting things quickly. Since the goal of this book is not to teach you all about R but just how to produce good graphics, we can postpone a lot of details until later (or indeed ignore them indefinitely). We will start as we mean to go on, by using a function to make a named object, and plot the result.

We will use the Gapminder dataset, which you should already have available on your computer. We load the data with `library()` and take a look.

```
library(gapminder)
gapminder
```

```
## # A tibble: 1,704 x 6
##    country     continent  year lifeExp      pop gdpPercap
##    <fct>       <fct>     <int>   <dbl>    <int>     <dbl>
##  1 Afghanistan Asia       1952    28.8  8425333      779.
##  2 Afghanistan Asia       1957    30.3  9240934      821.
##  3 Afghanistan Asia       1962    32.0 10267083      853.
##  4 Afghanistan Asia       1967    34.0 11537966      836.
##  5 Afghanistan Asia       1972    36.1 13079460      740.
##  6 Afghanistan Asia       1977    38.4 14880372      786.
##  7 Afghanistan Asia       1982    39.9 12881816      978.
##  8 Afghanistan Asia       1987    40.8 13867957      852.
##  9 Afghanistan Asia       1992    41.7 16317921      649.
## 10 Afghanistan Asia       1997    41.8 22227415      635.
## # ... with 1,694 more rows
```

This is a table of data about a large number of countries, each observed over several years. Let's make a scatterplot with it. Type the code below and try to get a sense of what's happening. Don't worry too much yet about the details.

```
p ← ggplot(data = gapminder,
           mapping = aes(x = gdpPercap, y = lifeExp))
p + geom_point()
```

Not a bad start. Our graph (fig. 2.6) is fairly legible, it has its axes informatively labeled, and it shows some sort of relationship between the two variables we have chosen. It could also be made better. Let's learn more about how to improve it.

2.7 Where to Go Next

You should go straight to the next chapter. However, you could also spend a little more time getting familiar with R and RStudio.

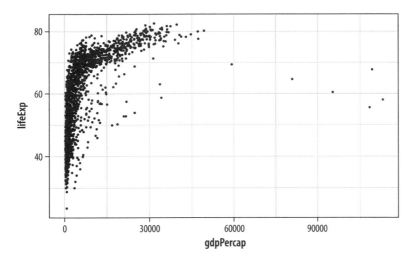

Figure 2.6: Life expectancy plotted against GDP per capita for a large number of country-years.

Some information in the appendix to this book might already be worth glancing at, especially the additional introductory material on R, and the discussion there about some common problems that tend to happen when reading in your own data. There are several free or initially free online introductions to the R language that are worth trying. You do not need to know the material they cover in order to keep going with this book, but you might find one or more of them useful. If you get a little bogged down in any of them or find the examples they choose are not that relevant to you, don't worry. These introductions tend to want to introduce you to a range of programming concepts and tools that we will not need right away.

swirlstats.com
tryr.codeschool.com
datacamp.com

It is also worth familiarizing yourself a little with how RStudio works and with what it can do for you. The RStudio website has a great deal of introductory material to help you along. You can also find a number of handy cheat sheets there that summarize different pieces of RStudio, RMarkdown, and various tidyverse packages that we will use throughout the book. These cheat sheets are not meant to teach you the material, but they are helpful points of reference once you are up and running.

rstudio.com

rstudio.com/resources/cheatsheets

3 Make a Plot

This chapter will teach you how to use ggplot's core functions to produce a series of scatterplots. From one point of view, we will proceed slowly and carefully, taking our time to understand the logic behind the commands that you type. The reason for this is that the central activity of visualizing data with ggplot more or less *always* involves the same sequence of steps. So it is worth learning what they are.

From another point of view, though, we will move fast. Once you have the basic sequence down and understand how it is that ggplot assembles the pieces of a plot into a final image, then you will find that analytically and aesthetically sophisticated plots come within your reach very quickly. By the end of this chapter, for example, we will have learned how to produce a small-multiple plot of time-series data for a large number of countries, with a smoothed regression line in each panel.

3.1 How Ggplot Works

As we saw in chapter 1, visualization involves representing your data using lines or shapes or colors and so on. There is some structured relationship, some *mapping*, between the variables in your data and their representation in the plot displayed on your screen or on the page. We also saw that not all mappings make sense for all types of variables, and (independently of this) some representations are harder to interpret than others. Ggplot provides you with a set of tools to map data to visual elements on your plot, to specify the kind of plot you want, and then to control the fine details of how it will be displayed. Figure 3.1 shows a schematic outline of the process starting from data, at the top, down to a finished plot at the bottom. Don't worry about the details for now. We will be going into them one piece at a time over the next few chapters.

The most important thing to get used to with ggplot is the way you use it to think about the logical structure of your plot. The code you write specifies the connections between the variables in your data, and the colors, points, and shapes you see on the screen. In ggplot, these logical connections between your data and the plot elements are called *aesthetic mappings* or just *aesthetics*. You begin every plot by telling the ggplot() function what your data is and how the variables in this data logically map onto the plot's aesthetics. Then you take the result and say what general sort of plot you want, such as a scatterplot, a boxplot, or a bar chart. In ggplot, the overall type of plot is called a *geom*. Each geom has a function that creates it. For example, geom_point() makes scatterplots, geom_bar() makes bar plots, geom_boxplot() makes boxplots, and so on. You combine these two pieces, the ggplot() object and the geom, by literally adding them together in an expression, using the "+" symbol.

At this point, ggplot will have enough information to be able to draw a plot for you. The rest is just details about exactly what you want to see. If you don't specify anything further, ggplot will use a set of defaults that try to be sensible about what gets drawn. But more often you will want to specify exactly what you want, including information about the scales, the labels of legends and axes, and other guides that help people to read the plot. These pieces are added to the plot in the same way as the geom_ function was. Each component has its own function, you provide arguments to it specifying what to do, and you literally add it to the sequence of instructions. In this way you systematically build your plot piece by piece.

In this chapter we will go through the main steps of this process. We will proceed by example, repeatedly building a series of plots. As noted earlier, I *strongly* encourage you go through this exercise manually, typing (rather than copying and pasting) the code yourself. This may seem a bit tedious, but it is *by far* the most effective way to get used to what is happening, and to get a feel for R's syntax. While you'll inevitably make some errors, you will also quickly find yourself becoming able to diagnose your own mistakes, as well as having a better grasp of the higher-level structure of plots. You should open the RMarkdown file for your notes, remember to load the tidyverse package, and write the code out in chunks, interspersing your own notes and comments as you go.

```
library(tidyverse)
```

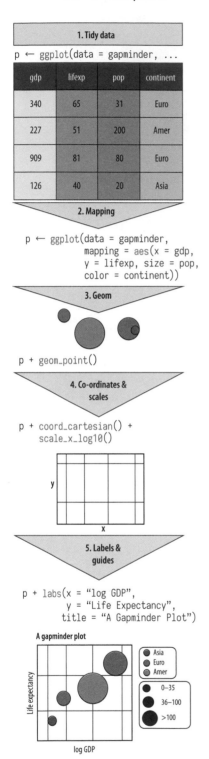

Figure 3.1: The main elements of ggplot's grammar of graphics. This chapter goes through these steps in detail.

3.2 Tidy Data

The tidyverse tools we will be using want to see your data in a particular sort of shape, generally referred to as "tidy data" (Wickham 2014). Social scientists will likely be familiar with the distinction between *wide-format* and *long-format* data. In a long-format table, every variable is a column, and every observation is a row. In a wide-format table, some variables are spread out across columns. For example, table 3.1 shows part of a table of life expectancy over time for a series of countries. This is in wide format because one of the variables, year, is spread across the columns of the table.

By contrast, table 3.2 shows the beginning of the same data in long format. The tidy data that ggplot wants is in this long form. In a related bit of terminology, in this table the year variable is sometimes called a *key* and the lifeExp variable is the *value* taken by that key for any particular row. These terms are useful when converting tables from wide to long format. I am speaking fairly loosely here. Underneath these terms there is a worked-out theory of the forms that tabular data can be stored in, but right now we don't need to know those additional details. For more on the ideas behind tidy data, see the discussion in the appendix. You will also find an example showing the R code you need to get from an untidy to a tidy shape for the common "wide" case where some variables are spread out across the columns of your table.

If you compare tables 3.1 and 3.2, it is clear that a tidy table does not present data in its most compact form. In fact, it is usually not how you would choose to present your data if you wanted just to show people the numbers. Neither is untidy data "messy" or the "wrong" way to lay out data in some generic sense. It's just that, even if its long-form shape makes tables larger, tidy data is much more straightforward to work with when it comes to specifying the mappings that you need to coherently describe plots.

3.3 Mappings Link Data to Things You See

It's useful to think of a recipe or template that we start from each time we want to make a plot. This is shown in figure 3.2. We start with just one object of our own, our data, which should be in a

TABLE 3.1
Life Expectancy data in wide format.

country	1952	1957	1962	1967	1972	1977	1982	1987	1992	1997	2002	2007
Afghanistan	29	30	32	34	36	38	40	41	42	42	42	44
Albania	55	59	65	66	68	69	70	72	72	73	76	76
Algeria	43	46	48	51	55	58	61	66	68	69	71	72
Angola	30	32	34	36	38	39	40	40	41	41	41	43
Argentina	62	64	65	66	67	68	70	71	72	73	74	75
Australia	69	70	71	71	72	73	75	76	78	79	80	81

shape that ggplot understands. Usually this will be a *data frame* or some augmented version of it, like a *tibble*. We tell the core ggplot function what our data is. In this book, we will do this by creating an object named p, which will contain the core information for our plot. (The name p is just a convenience.) Then we choose a plot type, or *geom*, and add it to p. From there we add more features to the plot as needed, such as additional elements, adjusted scales, a title, or other labels.

We'll use the gapminder data to make our first plots. Make sure the library containing the data is loaded. If you are following through from the previous chapter in the same RStudio session or RMarkdown document, you won't have to load it again. Otherwise, use library() to make it available.

```
library(gapminder)
```

We can remind ourselves again what it looks like by typing the name of the object at the console:

```
gapminder
```

```
## # A tibble: 1,704 x 6
##    country     continent year lifeExp      pop gdpPercap
##    <fct>       <fct>     <int>   <dbl>    <int>     <dbl>
## 1 Afghanistan Asia       1952    28.8  8425333      779.
## 2 Afghanistan Asia       1957    30.3  9240934      821.
## 3 Afghanistan Asia       1962    32.0 10267083      853.
## 4 Afghanistan Asia       1967    34.0 11537966      836.
## 5 Afghanistan Asia       1972    36.1 13079460      740.
```

TABLE 3.2
Life Expectancy data in long format.

country	year	lifeExp
Afghanistan	1952	29
Afghanistan	1957	30
Afghanistan	1962	32
Afghanistan	1967	34
Afghanistan	1972	36
Afghanistan	1977	38

```
p ← ggplot(data= <data>,
       mapping= aes(<aesthetic> = <variable>,
                    <aesthetic> = <variable>,
                    <...> = <...>)

p + geom_<type>(<...>)+
    scale_<mapping>_<type>(<...>)+
    coord_<type>(<...>)+
    labs(<...>)
```

Figure 3.2: A schematic for making a plot.

```
## 6 Afghanistan Asia    1977    38.4 14880372    786.
## 7 Afghanistan Asia    1982    39.9 12881816    978.
## 8 Afghanistan Asia    1987    40.8 13867957    852.
## 9 Afghanistan Asia    1992    41.7 16317921    649.
## 10 Afghanistan Asia   1997    41.8 22227415    635.
## # ... with 1,694 more rows
```

Let's say we want to plot life expectancy against per capita GDP for all country-years in the data. We'll do this by creating an object that has some of the necessary information in it and build it up from there. First, we must tell the ggplot() function what data we are using.

Remember, use Option+minus on MacOS or Alt+minus on Windows to type the assignment operator.

```
p ← ggplot(data = gapminder)
```

At this point ggplot knows our data but not the *mapping*. That is, we need to tell it which variables in the data should be represented by which visual elements in the plot. It also doesn't know what sort of plot we want. In ggplot, mappings are specified using the aes() function, like this:

You do not need to explicitly name the arguments you pass to functions, as long as you provide them in the expected order, viz, the order listed on the help page for the function. This code would still work if we omitted data = and mapping = . In this book, I name all the arguments for clarity.

```
p ← ggplot(data = gapminder, mapping = aes(x = gdpPercap, y = lifeExp))
```

Here we've given the ggplot() function two arguments instead of one: data and mapping. The data argument tells ggplot where to find the variables it is about to use. This saves us from having to tediously dig out the name of each variable in full. Instead, any mentions of variables will be looked for here first.

Next, the mapping. The mapping argument is not a data object, nor is it a character string. Instead, it's a function. (Remember, functions can be nested inside other functions.) The arguments we give to the aes function are a sequence of definitions that ggplot will use later. Here they say, "The variable on the x-axis is going to be gdpPercap, and the variable on the y-axis is going to be lifeExp." The aes() function does not say where variables with those names are to be found. That's because ggplot() is going to assume that things with that name are in the object given to the data argument.

The mapping = aes(...) argument *links variables* to *things you will see* on the plot. The x and y values are the most obvious ones. Other aesthetic mappings can include, for example, color, shape, size, and line type (whether a line is solid, dashed, or some other pattern). We'll see examples in a minute. A mapping does not directly say what particular colors or shapes, for example, will be on the plot. Rather it says which *variables* in the data will be *represented* by visual elements like a color, a shape, or a point on the plot area.

What happens if we just type p at the console at this point and hit return? The result is shown in figure 3.3.

```
p
```

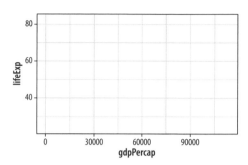

Figure 3.3: This empty plot has no geoms.

The p object has been created by the ggplot() function and already has information in it about the mappings we want, together with a lot of other information added by default. (If you want to see just how much information is in the p object already, try asking for str(p).) However, we haven't given it any instructions yet about what sort of plot to draw. We need to add a *layer* to the plot. This means picking a geom_ function. We will use geom_point(). It knows how to take x and y values and plot them in a scatterplot.

```
p + geom_point()
```

3.4 Build Your Plots Layer by Layer

Although we got a brief taste of ggplot at the end of chapter 2, we spent more time in that chaper preparing the ground to make this first proper graph. We set up our software IDE and made sure we

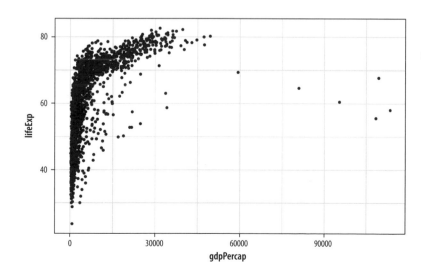

Figure 3.4: A scatterplot of life expectancy vs GDP.

could reproduce our work. We then learned the basics of how R works, and the sort of tidy data that ggplot expects. Just now we went through the logic of ggplot's main idea, of building up plots a piece at a time in a systematic and predictable fashion, beginning with a mapping between a variable and an aesthetic element. We have done a lot of work and produced one plot.

The good news is that, from now on, not much will change conceptually about what we are doing. It will be more a question of learning in greater detail about how to tell ggplot what to do. We will learn more about the different geoms (or types of plot) available and find out about the functions that control the coordinate system, scales, guiding elements (like labels and tick-marks), and thematic features of plots. This will allow us to make much more sophisticated plots surprisingly fast. Conceptually, however, we will always be doing the same thing. We will start with a table of data that has been tidied, and then we will do the following:

The data = ... step.

The mapping = aes(...) step.

1. Tell the ggplot() function what our data is.
2. Tell ggplot() *what* relationships we want to see. For convenience we will put the results of the first two steps in an object called p.

Choose a geom.

3. Tell ggplot *how* we want to see the relationships in our data.
4. Layer on geoms as needed, by adding them to the p object one at a time.

The scale_, family, labs() and guides() functions.

5. Use some additional functions to adjust scales, labels, tick-marks, titles. We'll learn more about some of these functions shortly.

To begin with we will let ggplot use its defaults for many of these elements. The coordinate system used in plots is most often cartesian, for example. It is a plane defined by an x-axis and a y-axis. This is what ggplot assumes, unless you tell it otherwise. But we will quickly start making some adjustments. Bear in mind once again that the process of adding layers to the plot really is *additive*. Usually in R, functions cannot simply be added to objects. Rather, they take objects as inputs and produce objects as outputs. But the objects created by ggplot() are special. This makes it easier to assemble plots one piece at a time, and to inspect how they look at every step. For example, let's try a different geom_ function with our plot.

In effect we create one big object that is a nested list of instructions for how to draw each piece of the plot.

```
p ← ggplot(data = gapminder, mapping = aes(x = gdpPercap, y = lifeExp))
p + geom_smooth()
```

You can see right away in figure 3.5 that some of these geoms do a lot more than simply put points on a grid. Here geom_smooth() has calculated a smoothed line for us and shaded in a ribbon showing the standard error for the line. If we want to see the data points and the line together (fig. 3.6), we simply add geom_point() back in:

```
p ← ggplot(data = gapminder, mapping = aes(x = gdpPercap, y = lifeExp))
p + geom_point() + geom_smooth()
```

```
## `geom_smooth()` using method = 'gam' and formula 'y ~ s(x, bs = "cs")'
```

The console message R tells you the geom_smooth() function is using a method called gam, which in this case means it has fit a generalized additive model. This suggests that maybe there are other methods that geom_smooth() understands, and which we might tell it to use instead. Instructions are given to functions via their arguments, so we can try adding method = "lm" (for "linear model") as an argument to geom_smooth():

```
p ← ggplot(data = gapminder, mapping = aes(x = gdpPercap, y = lifeExp))
p + geom_point() + geom_smooth(method = "lm")
```

For figures 3.5 to 3.7 we did not have to tell geom_point() or geom_smooth() where their data was coming from, or what mappings they should use. They *inherit* this information from the original p object. As we'll see later, it's possible to give geoms separate instructions that they will follow instead. But in the absence of any other information, the geoms will look for the instructions they need in the ggplot() function, or the object created by it.

In our plot, the data is quite bunched up against the left side. Gross domestic product per capita is not normally distributed across our country years. The x-axis scale would probably look better if it were transformed from a linear scale to a log scale. For this we can use a function called scale_x_log10(). As you might expect, this function scales the x-axis of a plot to a log 10 basis. To use it we just add it to the plot:

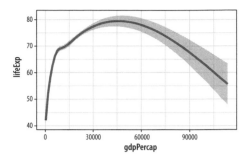

Figure 3.5: Life expectancy vs GDP, using a smoother.

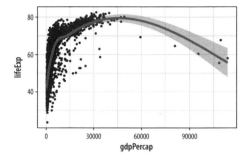

Figure 3.6: Life expectancy vs GDP, showing both points and a GAM smoother.

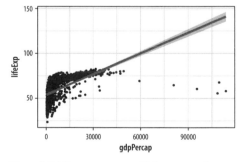

Figure 3.7: Life expectancy vs GDP, points and an ill-advised linear fit.

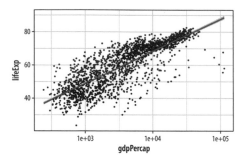

Figure 3.8: Life expectancy vs GDP scatterplot, with a GAM smoother and a log scale on the x-axis.

```
p ← ggplot(data = gapminder, mapping = aes(x = gdpPercap, y = lifeExp))
p + geom_point() + geom_smooth(method = "gam") + scale_x_log10()
```

The x-axis transformation repositions the points and also changes the shape the smoothed line (fig. 3.8). (We switched back to gam from lm.) While ggplot() and its associated functions have not made any changes to our underlying data frame, the scale transformation is applied to the data before the smoother is layered onto the plot. There are a variety of scale transformations that you can use in just this way. Each is named for the transformation you want to apply, and the axis you want to apply it to. In this case we use scale_x_log10().

At this point, if our goal was just to show a plot of life expectancy vs GDP using sensible scales and adding a smoother, we would be thinking about polishing up the plot with nicer axis labels and a title. Perhaps we might also want to replace the scientific notation on the x-axis with the dollar value it actually represents. We can do both of these things quite easily. Let's take care of the scale first. The labels on the tick-marks can be controlled through the scale_ functions. While it's possible to roll your own function to label axes (or just supply your labels manually, as we will see later), there's also a handy scales package that contains some useful premade formatting functions. We can either load the whole package with library(scales) or, more conveniently, just grab the specific formatter we want from that library. Here it's the dollar() function. To grab a function directly from a package we have not loaded, we use the syntax thepackage::thefunction. So we can do this:

```
p ← ggplot(data = gapminder, mapping = aes(x = gdpPercap, y = lifeExp))
p + geom_point() +
    geom_smooth(method = "gam") +
    scale_x_log10(labels = scales::dollar)
```

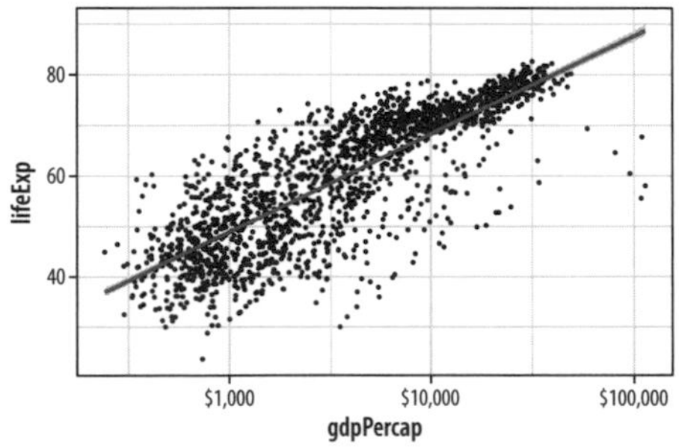

We will learn more about scale transformations later. For now, just remember two things about them. First, you can directly transform your x- or y-axis by adding something like scale_x_log10() or scale_y_log10() to your plot. When you do so, the x- or y-axis will be transformed, and, by default, the tick-marks on the axis will be labeled using scientific notation. Second, you can give these scale_ functions a labels argument that reformats

Figure 3.9: Life expectancy vs GDP scatterplot, with a GAM smoother and a log scale on the x-axis, with better labels on the tick-marks.

the text printed underneath the tick-marks on the axes. Inside the scale_x_log10() function try labels=scales::comma, for example.

3.5 Mapping Aesthetics vs Setting Them

An *aesthetic mapping* specifies that a variable will be expressed by one of the available visual elements, such as size, or color, or shape. As we've seen, we map variables to aesthetics like this:

```
p ← ggplot(data = gapminder, mapping = aes(x = gdpPercap, y - lifeExp,
    color = continent))
```

This code does *not* give a direct instruction like "color the points purple." Instead it says, "the property 'color' will represent the variable continent," or "color will map continent." If we want to turn all the points in the figure purple, we do *not* do it through the mapping function. Look at what happens when we try:

```
p ← ggplot(data = gapminder, mapping = aes(x = gdpPercap, y = lifeExp,
    color = "purple"))
p + geom_point() + geom_smooth(method = "loess") + scale_x_log10()
```

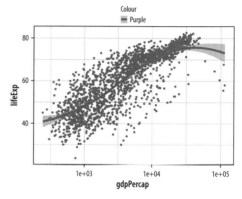

Figure 3.10: What has gone wrong here?

What has happened in figure 3.10? Why is there a legend saying "purple"? And why have the points all turned pinkish-red instead of purple? Remember, an aesthetic is a mapping of variables in your data to properties you can see on the graph. The aes() function is where that mapping is specified, and the function is trying to do its job. It wants to map a variable to the color aesthetic, so it assumes you are giving it a variable. We have only given it one word, though—"purple." Still, aes() will do its best to treat that word as though it were a variable. A variable should have as many observations as there are rows in the data, so aes() falls back on R's recycling rules for making vectors of different lengths match up.

In effect, this creates a new categorical variable for your data. The string "purple" is recycled for every row of your data. Now you have a new column. Every element in it has the same value, "purple." Then ggplot plots the results on the graph as you've asked it to, by mapping it to the color aesthetic. It dutifully makes a legend for this new variable. By default, ggplot displays the points falling

Just as in chapter 1, when we were able to write 'my_numbers + 1' to add one to each element of the vector.

into the category "purple" (which is all of them) using its default first-category hue, which is red.

The aes() function is for mappings only. Do not use it to change properties to a particular value. If we want to *set* a property, we do it in the geom_ we are using, and *outside* the mapping = aes(...) step. Try this:

```
p ← ggplot(data = gapminder, mapping = aes(x = gdpPercap, y = lifeExp))
p + geom_point(color = "purple") + geom_smooth(method = "loess") +
    scale_x_log10()
```

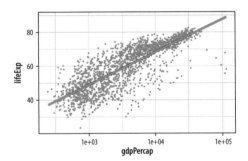

Figure 3.11: Setting the color attribute of the points directly.

The geom_point() function can take a color argument directly, and R knows what color "purple" is (fig. 3.11). This is not part of the aesthetic mapping that defines the basic structure of the graphic. From the point of view of the grammar or logic of the graph, the fact that the points are colored purple has no significance. The color purple is not representing or mapping a variable or feature of the data in the relevant way.

```
p ← ggplot(data = gapminder, mapping = aes(x = gdpPercap, y = lifeExp))
p + geom_point(alpha = 0.3) + geom_smooth(color = "orange", se = FALSE,
    size = 8, method = "lm") + scale_x_log10()
```

Figure 3.12: Setting some other arguments.

It's also possible to map a continuous variable directly to the alpha property, much like one might map a continuous variable to a single-color gradient. However, this is generally not an effective way of precisely conveying variation in quantity.

The various geom_ functions can take many other arguments that will affect how the plot looks but do not involve mapping variables to aesthetic elements. Thus those arguments will never go inside the aes() function. Some of the things we will want to set, like color or size, have the same name as mappable elements. Others, like the method or se arguments in geom_smooth() affect other aspects of the plot. In the code for figure 3.12, the geom_smooth(call sets the line color to orange and sets its size (i.e., thickness) to 8, an unreasonably large value. We also turn off the se option by switching it from its default value of TRUE to FALSE. The result is that the standard error ribbon is not shown.

Meanwhile in the geom_smooth() call we set the alpha argument to 0.3. Like color, size, and shape, "alpha" is an aesthetic property that points (and some other plot elements) have, and to which variables can be mapped. It controls how transparent the object will appear when drawn. It's measured on a scale of zero to one. An object with an alpha of zero will be completely transparent. Setting it to zero will make any other mappings the object might have, such as color or size, invisible as well. An object with an alpha of one will be completely opaque. Choosing an intermediate value can be useful when there is a lot of overlapping data to plot (as in fig. 3.13), as it makes it easier to see where the bulk of the observations are located.

```
p ← ggplot(data = gapminder, mapping = aes(x = gdpPercap, y=lifeExp))
p + geom_point(alpha = 0.3) +
    geom_smooth(method = "gam") +
    scale_x_log10(labels = scales::dollar) +
    labs(x = "GDP Per Capita", y = "Life Expectancy in Years",
        title = "Economic Growth and Life Expectancy",
        subtitle = "Data points are country-years",
        caption = "Source: Gapminder.")
```

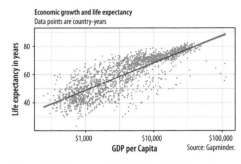

Figure 3.13: A more polished plot of Life Expectancy vs GDP.

We can now make a reasonably polished plot. We set the alpha of the points to a low value, make nicer x- and y-axis labels, and add a title, subtitle, and caption. As you can see in the code above, in addition to x, y, and any other aesthetic mappings in your plot (such as size, fill, or color), the labs() function can also set the text for title, subtitle, and caption. It controls the *main labels* of scales. The appearance of things like axis tickmarks is the responsibility of various scale_ functions, such as the scale_x_log10() function used here. We will learn more about what can be done with scale_ functions soon.

Are there any variables in our data that can sensibly be mapped to the color aesthetic? Consider continent. In figure 3.14 the individual data points have been colored by continent, and a legend with a key to the colors has automatically been added to the plot. In addition, instead of one smoothing line we now have five. There is one for each unique value of the continent variable. This is a consequence of the way aesthetic mappings are inherited. Along with x and y, the color aesthetic mapping is set in the call to ggplot() that we used to create the p object. Unless told otherwise, all geoms layered on top of the original plot object will inherit that object's mappings. In this case we get both our points and smoothers colored by continent.

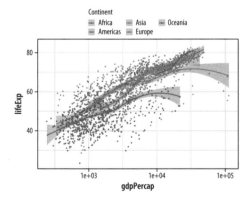

Figure 3.14: Mapping the continent variable to the color aesthetic.

```
p ← ggplot(data = gapminder, mapping = aes(x = gdpPercap, y = lifeExp,
    color = continent))
p + geom_point() + geom_smooth(method = "loess") + scale_x_log10()
```

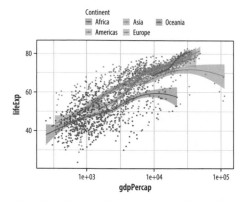

Figure 3.15: Mapping the continent variable to the color aesthetic, and correcting the error bars using the fill aesthetic.

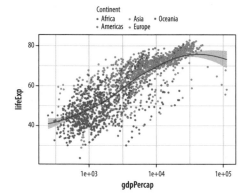

Figure 3.16: Mapping aesthetics on a per geom basis. Here color is mapped to continent for the points but not the smoother.

If it is what we want, then we might also consider shading the standard error ribbon of each line to match its dominant color, as in figure 3.15. The color of the standard error ribbon is controlled by the fill aesthetic. Whereas the color aesthetic affects the appearance of lines and points, fill is for the filled areas of bars, polygons and, in this case, the interior of the smoother's standard error ribbon.

```
p ← ggplot(data = gapminder, mapping = aes(x = gdpPercap, y = lifeExp,
    color = continent, fill = continent))
p + geom_point() + geom_smooth(method = "loess") + scale_x_log10()
```

Making sure that color and fill aesthetics match up consistently in this way improves the overall look of the plot. To make it happen we just need to specify that the mappings are to the same variable in each case.

3.6 Aesthetics Can Be Mapped per Geom

Perhaps five separate smoothers is too many, and we just want one line. But we still would like to have the points color-coded by continent. By default, geoms inherit their mappings from the ggplot() function. We can change this by specifying different aesthetics for each geom. We use the same mapping = aes(...) expression as in the initial call to ggplot() but now use it in the geom_ functions as well, specifying the mappings we want to apply to each one (fig. 3.16). Mappings specified in the initial ggplot() function—here, x and y—will carry through to all subsequent geoms.

```
p ← ggplot(data = gapminder, mapping = aes(x = gdpPercap, y = lifeExp))
p + geom_point(mapping = aes(color = continent)) +
    geom_smooth(method = "loess") +
    scale_x_log10()
```

It's possible to map continuous variables to the color aesthetic, too. For example, we can map the log of each country-year's population (pop) to color. (We can take the log of population right in the aes() statement, using the log() function. R will evaluate this for us.) When we do this, ggplot produces a gradient scale. It is continuous but is marked at intervals in the legend. Depending on the circumstances, mapping quantities like population to a continuous color gradient (fig. 3.17) may be more or less effective than cutting the variable into categorical bins running, e.g., from low to high. In general it is always worth looking at the data in its continuous form first rather than cutting or binning it into categories.

```
p ← ggplot(data = gapminder, mapping = aes(x = gdpPercap, y = lifeExp))
p + geom_point(mapping = aes(color = log(pop))) + scale_x_log10()
```

Finally, it is worth paying a little more attention to the way that ggplot draws its scales. Because every mapped variable has a scale, we can learn a lot about how a plot has been constructed, and what mappings it contains, by seeing what the legends look like. For example, take a closer look at the legends produced in figures 3.15 and 3.16.

In the legend for the first figure, shown in figure 3.18 on the left, we see several visual elements. The key for each continent shows a dot, a line, and a shaded background. The key for the second figure, shown on the right, has only a dot for each continent, with no shaded background or line. If you look again at the code for figures 3.15 and 3.16, you will see that in the first case we mapped the continent variable to both color and fill. We then drew the figure with geom_point() and fitted a line for each continent with geom_smooth(). Points have color but the smoother understands both color (for the line itself) and fill (for the shaded standard error ribbon). Each of these elements is represented in the legend: the point color, the line color, and the ribbon fill. In the second figure, we decided to simplify things by having only the points be colored by continent. Then we drew just a single smoother for the whole graph. Thus, in the legend for that figure, the colored line and the shaded box are both absent. We see only a legend for the mapping of color to continent in geom_point(). Meanwhile on the graph itself the line drawn by geom_smooth() is set by default to a bright blue, different from anything on the scale, and its shaded error ribbon is set by default to gray. Small details like this are not accidents. They are a direct consequence of ggplot's grammatical way of thinking about the relationship between the data behind the plot and the visual elements that represent it.

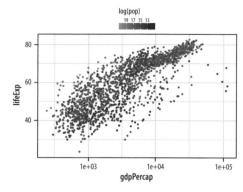

Figure 3.17: Mapping a continuous variable to color.

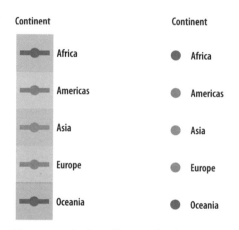

Figure 3.18: Guides and legends faithfully reflect the mappings they represent.

3.7 Save Your Work

Now that you have started to make your own plots, you may be wondering how to save them, and perhaps also how to control their size and format. If you are working in an RMarkdown document, then the plots you make will be embedded in it, as we have already seen. You can set the default size of plots within your .Rmd document by setting an option in your first code chunk. This one tells R to make 8 x 5 figures:

```
knitr::opts_chunk$set(fig.width = 8, fig.height = 5)
```

Because you will be making plots of different sizes and shapes, sometimes you will want to control the size of particular plots, without changing the default. To do this, you can add the same options to any particular chunk inside the curly braces at the beginning. Remember, each chunk opens with three backticks and then a pair of braces containing the language name (for us always r) and an optional label:

```
```{r example}
p + geom_point()
```
```

You can follow the label with a comma and provide a series of options if needed. They will apply only to that chunk. To make a figure twelve inches wide and nine inches high we say, e.g., {r example, fig.width = 12, fig.height = 9} in the braces section.

You will often need to save your figures individually, as they will end up being dropped into slides or published in papers that are not produced using RMarkdown. Saving a figure to a file can be done in several different ways. When working with ggplot, the easiest way is to use the ggsave() function. To save the most recently displayed figure, we provide the name we want to save it under:

```
ggsave(filename = "my_figure.png")
```

This will save the figure as a PNG file, a format suitable for displaying on web pages. If you want a PDF instead, change the extension of the file:

Several other file formats are available as well. See the function's help page for details.

```
ggsave(filename = "my_figure.pdf")
```

Remember that, for convenience, you do not need to write filename = as long as the name of the file is the first argument you give ggsave(). You can also pass plot objects to ggsave(). For example, we can put our recent plot into an object called p_out and then tell ggsave() that we want to save that object.

```
p_out ← p + geom_point() + geom_smooth(method = "loess") + scale_x_log10()

ggsave("my_figure.pdf", plot = p_out)
```

When saving your work, it is useful to have one or more subfolders where you save only figures. You should also take care to name your saved figures in a sensible way. fig_1.pdf or my_figure.pdf are not good names. Figure names should be compact but descriptive, and consistent between figures within a project. In addition, although it really shouldn't be the case in this day and age, it is also wise to play it safe and avoid file names containing characters likely to make your code choke in future. These include apostrophes, backticks, spaces, forward and back slashes, and quotes.

The appendix contains a short discussion of how to organize your files within your project folder. Treat the project folder as the home base of your work for the paper or work you are doing, and put your data and figures in subfolders within the project folder. To begin with, using your file manager, create a folder named "figures" inside your project folder. When saving figures, you can use Kirill Müller's handy here library to make it easier to work with files and subfolders while not having to type in full file paths. Load the library in the setup chunk of your RMarkdown document. When you do, it tells you where "here" is for the current project. You will see a message saying something like this, with your file path and user name instead of mine:

```
# here() starts at /Users/kjhealy/projects/socviz
```

You can then use the here() function to make loading and saving your work more straightforward and safer. Assuming a folder named "figures" exists in your project folder, you can do this:

```
ggsave(here("figures", "lifexp_vs_gdp_gradient.pdf"), plot = p_out)
```

This saves p_out as a file called lifeexp_vs_gdp_gradient.pdf in the figures directory *here*, i.e., in your current project folder.

You can save your figure in a variety of formats, depending on your needs (and also, to a lesser extent, on your particular computer system). The most important distinction to bear in mind is between *vector* formats and *raster* formats. A file with a *vector* format, like PDF or SVG, is stored as a set of instructions about lines, shapes, colors, and their relationships. The viewing software (such as Adobe Acrobat or Apple's Preview application for PDFs) then interprets those instructions and displays the figure. Representing the figure this way allows it to be easily resized without becoming distorted. The underlying language of the PDF format is Postscript, which is also the language of modern typesetting and printing. This makes a vector-based format like PDF the best choice for submission to journals.

A *raster*-based format, on the other hand, stores images essentially as a grid of pixels of a predefined size with information about the location, color, brightness, and so on of each pixel in the grid. This makes for more efficient storage, especially when used in conjunction with compression methods that take advantage of redundancy in images in order to save space. Formats like JPG are compressed raster formats. A PNG file is a raster image format that supports lossless compression. For graphs containing an awful lot of data, PNG files will tend to be much smaller than the corresponding PDF. However, raster formats cannot be easily resized. In particular they cannot be expanded in size without becoming pixelated or grainy. Formats like JPG and PNG are the standard way that images are displayed on the web. The more recent SVG format is vector-based format but also nevertheless supported by many web browsers.

In general you should save your work in several different formats. When you save in different formats and in different sizes you may need to experiment with the scaling of the plot and the size of the fonts in order to get a good result. The scale argument to ggsave() can help you here (you can try out different values, like scale=1.3, scale=5, and so on). You can also use ggsave() to explicitly set the height and width of your plot in the units that you choose.

```
ggsave(here("figures", "lifexp_vs_gdp_gradient.pdf"), plot = p_out,
    height = 8, width = 10, units = "in")
```

Now that you know how to do that, let's get back to making more graphs.

3.8 Where to Go Next

Start by playing around with the gapminder data a little more. You can try each of these explorations with geom_point() and then with geom_smooth(), or both together.

- What happens when you put the geom_smooth() function before geom_point() instead of after it? What does this tell you about how the plot is drawn? Think about how this might be useful when drawing plots.
- Change the mappings in the aes() function so that you plot life expectancy against population (pop) rather than per capita GDP. What does that look like? What does it tell you about the unit of observation in the dataset?
- Try some alternative scale mappings. Besides scale_x_log 10(), you can try scale_x_sqrt() and scale_x_reverse(). There are corresponding functions for y-axis transformations. Just write y instead of x. Experiment with them to see what effect they have on the plot, and whether they make any sense to use.
- What happens if you map color to year instead of continent? Is the result what you expected? Think about what class of object year is. Remember you can get a quick look at the top of the data, which includes some shorthand information on the class of each variable, by typing gapminder.
- Instead of mapping color = year, what happens if you try color = factor(year)?
- As you look at these different scatterplots, think about figure 3.13 a little more critically. We worked it up to the point where it was reasonably polished, but is it really the best way to

display this country-year data? What are we gaining and losing by ignoring the temporal and country-level structure of the data? How could we do better? Sketch out what an alternative visualization might look like.

As you begin to experiment, remember two things. First, it's always worth trying something, even if you're not sure what's going to happen. Don't be afraid of the console. The nice thing about making your graphics through code is that you won't break anything you can't reproduce. If something doesn't work, you can figure out what happened, fix things, and rerun the code to make the graph again.

Second, remember that the main flow of action in ggplot is always the same. You start with a table of data, you map the variables you want to display to aesthetics like position, color, or shape, and you choose one or more geoms to draw the graph. In your code this gets accomplished by making an object with the basic information about data and mappings, and then adding or layering additional information as needed. Once you get used to this way of thinking about your plots, especially the aesthetic mapping part, drawing them becomes easier. Instead of having to think about how to draw particular shapes or colors on the screen, the many geom_ functions take care of that for you. In the same way, learning new geoms is easier once you think of them as ways to display aesthetic mappings that you specify. Most of the learning curve with ggplot involves getting used to this way of thinking about your data and its representation in a plot. In the next chapter, we will flesh out these ideas a little more, cover some common ways plots go "wrong" (i.e., when they end up looking strange), and learn how to recognize and avoid those problems.

This license does not extend to, for example, overwriting or deleting your data by mistake. You should still manage your project responsibly, at a minimum keeping good backups. But within R and at the level of experimenting with graphs at the console, you have a lot of freedom.

4   Show the Right Numbers

This chapter will continue to develop your fluency with ggplot's central workflow while also expanding the range of things you can do with it. One of our goals is to learn how to make new kinds of graph. This means learning some new geoms, the functions that make particular kinds of plots. But we will also get a better sense of what ggplot is doing when it draws plots, and learn more about how to write code that prepares our data to be plotted.

Code almost never works properly the first time you write it. This is the main reason that, when learning a new language, it is important to type out the exercises and follow along manually. It gives you a much better sense of how the syntax of the language works, where you're likely to make errors, and what the computer does when that happens. Running into bugs and errors is frustrating, but it's also an opportunity to learn a bit more. Errors can be obscure, but they are usually not malicious or random. If something has gone wrong, you can find out why it happened.

In R and ggplot, errors in code can result in figures that don't look right. We have already seen the result of one of the most common problems, when an aesthetic is mistakenly set to a constant value instead of being mapped to a variable. In this chapter we will discuss some useful features of ggplot that also commonly cause trouble. They have to do with how to tell ggplot more about the internal structure of your data (*grouping*), how to break up your data into pieces for a plot (*faceting*), and how to get ggplot to perform some calculations on or summarize your data before producing the plot (*transforming*). Some of these tasks are part of ggplot proper, so we will learn more about how geoms, with the help of their associated *stat* functions, can act on data before plotting it. As we shall also see, while it is possible to do a lot of transformation directly in ggplot, there can be more convenient ways to approach the same task.

4.1 Colorless Green Data Sleeps Furiously

When you write ggplot code in R you are in effect trying to "say" something visually. It usually takes several iterations to say exactly what you mean. This is more than a metaphor here. The ggplot library is an implementation of the "grammar" of graphics, an idea developed by Wilkinson (2005). The grammar is a set of rules for producing graphics from data, taking pieces of data and mapping them to geometric objects (like points and lines) that have aesthetic attributes (like position, color, and size), together with further rules for transforming the data if needed (e.g., to a smoothed line), adjusting scales (e.g., to a log scale), and projecting the results onto a different coordinate system (usually Cartesian).

A key point is that, like other rules of syntax, the grammar limits the structure of what you can say, but it does not automatically make what you say sensible or meaningful. It allows you to produce long "sentences" that begin with mappings of data to visual elements and add clauses about what sort of plot it is, how the axes are scaled, and so on. But these sentences can easily be garbled. Sometimes your code will not produce a plot at all because of some syntax error in R. You will forget a + sign between geom_ functions or lose a parenthesis somewhere so that your function statement becomes unbalanced. In those cases R will complain (perhaps in an opaque way) that something has gone wrong. At other times, your code will successfully produce a plot, but it will not look the way you expected it to. Sometimes the results will look very weird indeed. In those cases, the chances are you have given ggplot a series of grammatically correct instructions that are either nonsensical in some way or have accidentally twisted what you meant to say. These problems often arise when ggplot does not have quite all the information it needs in order make your graphic say what you want it to say.

We will see some alternatives to Cartesian coordinates later.

4.2 Grouped Data and the "Group" Aesthetic

Let's begin again with our Gapminder dataset. Imagine we wanted to plot the trajectory of life expectancy over time for each country in the data. We map year to x and lifeExp to y. We take a quick look at the documentation and discover that geom_line() will draw lines by connecting observations in order of the variable on the x-axis, which seems right. We write our code:

```
p ← ggplot(data = gapminder, mapping = aes(x = year, y = gdpPercap))
p + geom_line()
```

Something has gone wrong in figure 4.1. What happened? While ggplot will make a pretty good guess as to the structure of the data, it does not know that the yearly observations in the data are grouped by country. We have to tell it. Because we have not, geom_line() gamely tries to join up all the lines for each particular year in the order they appear in the dataset, as promised. It starts with an observation for 1952 in the first row of the data. It doesn't know this belongs to Afghanistan. Instead of going to Afghanistan 1953, it finds there are a series of 1952 observations, so it joins all those up first, alphabetically by country, all the way down to the 1952 observation that belongs to Zimbabwe. Then it moves to the first observation in the next year, 1957.

The result is meaningless when plotted. Bizarre-looking output in ggplot is common enough because everyone works out their plots one bit at a time, and making mistakes is just a feature of puzzling out how you want the plot to look. When ggplot successfully makes a plot but the result looks insane, the reason is almost always that something has gone wrong in the mapping between the data and aesthetics for the geom being used. This is so common there's even a Twitter account devoted to the "Accidental aRt" that results. So don't despair!

In this case, we can use the group aesthetic to tell ggplot explicitly about this country-level structure.

```
p ← ggplot(data = gapminder, mapping = aes(x = year, y = gdpPercap))
p + geom_line(aes(group = country))
```

The plot in figure 4.2 is still fairly rough, but it is showing the data properly, with each line representing the trajectory of a country over time. The gigantic outlier is Kuwait, in case you are interested.

The group aesthetic is usually only needed when the grouping information you need to tell ggplot about is not built into the variables being mapped. For example, when we were plotting the

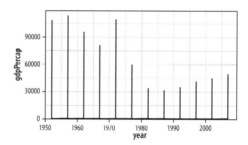

Figure 4.1: Trying to plot the data over time by country.

This would have worked if there were only one country in the dataset.

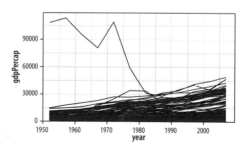

Figure 4.2: Plotting the data over time by country, again.

points by continent, mapping color to continent was enough to get the right answer because continent is already a categorical variable, so the grouping is clear. When mapping x to year, however, there is no information in the year variable itself to let ggplot know that it is grouped by country for the purposes of drawing lines with it. So we need to say that explicitly.

4.3 Facet to Make Small Multiples

The plot we just made has a lot of lines on it. While the overall trend is more or less clear, it looks a little messy. One option is to *facet* the data by some third variable, making a "small multiple" plot. This is a powerful technique that allows a lot of information to be presented compactly and in a consistently comparable way. A separate panel is drawn for each value of the faceting variable. Facets are not a geom but rather a way of organizing a series of geoms. In this case we have the continent variable available to us. We will use facet_wrap() to split our plot by continent.

The facet_wrap() function can take a series of arguments, but the most important is the first one, which is specified using R's "formula" syntax, which uses the tilde character, ~. Facets are usually a one-sided formula. Most of the time you will just want a single variable on the right side of the formula. But faceting is powerful enough to accommodate what are in effect the graphical equivalent of multiway contingency tables, if your data is complex enough to require that. For our first example, we will just use a single term in our formula, which is the variable we want the data broken up by: facet_wrap(~ continent).

```
p ← ggplot(data = gapminder, mapping = aes(x = year, y = gdpPercap))
p + geom_line(aes(group = country)) + facet_wrap(~continent)
```

Each facet is labeled at the top. The overall layout of figure 4.3 minimizes the duplication of axis labels and other scales. Remember, too, that we can still include other geoms as before, and they will be layered within each facet. We can also use the ncol argument to facet_wrap() to control the number of columns used to lay out the facets. Because we have only five continents, it might be worth seeing if we can fit them on a single row (which means we'll

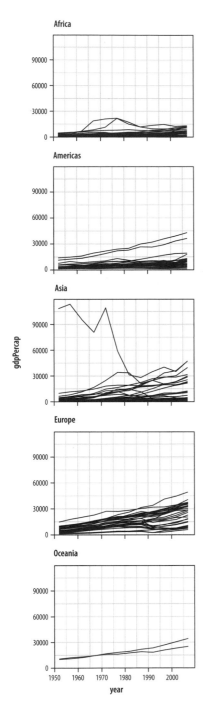

Figure 4.3: Faceting by continent.

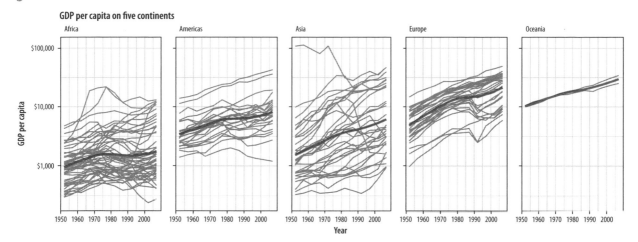

Figure 4.4: Faceting by continent, again.

have five columns). In addition, we can add a smoother, and a few cosmetic enhancements that make the graph a little more effective (fig. 4.4). In particular we will make the country trends a light gray color. We need to write a little more code to make all this happen. If you are unsure of what each piece of code does, take advantage of ggplot's additive character. Working backward from the bottom up, remove each + `some_function(...)` statement one at a time to see how the plot changes.

```
p ← ggplot(data = gapminder, mapping = aes(x = year, y = gdpPercap))
p + geom_line(color="gray70", aes(group = country)) +
    geom_smooth(size = 1.1, method = "loess", se = FALSE) +
    scale_y_log10(labels=scales::dollar) +
    facet_wrap(~ continent, ncol = 5) +
    labs(x = "Year",
         y = "GDP per capita",
         title = "GDP per capita on Five Continents")
```

This plot brings together an aesthetic mapping of x and y variables, a grouping aesthetic (`country`), two geoms (a lineplot and a smoother), a log-transformed y-axis with appropriate tick labels, a faceting variable (`continent`), and finally axis labels and a title.

The `facet_wrap()` function is best used when you want a series of small multiples based on a single categorical variable. Your panels will be laid out in order and then wrapped into a grid. If you wish you can specify the number of rows or the

We could also have faceted by country, which would have made the group mapping superfluous. But that would make almost 150 panels.

number of columns in the resulting layout. Facets can be more complex than this. For instance, you might want to cross-classify some data by two categorical variables. In that case you should try `facet_grid()` instead. This function will lay out your plot in a true two-dimensional arrangement, instead of a series of panels wrapped into a grid.

To see the difference, let's introduce `gss_sm`, a new dataset that we will use in the next few sections, as well as later on in the book. It is a small subset of the questions from the 2016 General Social Survey, or GSS. The GSS is a long-running survey of American adults that asks about a range of topics of interest to social scientists. The `gapminder` data consists mostly of *continuous* variables measured within countries by year. Measures like GDP per capita can take any value across a large range and they vary smoothly. The only *categorical* grouping variable is `continent`. It is an unordered categorical variable. Each country belongs to one continent, but the continents themselves have no natural ordering. By contrast, the GSS contains many categorical measures.

In social scientific work, especially when analyzing individual-level survey data, we often work with categorical data of various kinds. Sometimes the categories are unordered, as with ethnicity or sex. But they may also be ordered, as when we measure highest level of education attained on a scale ranging from elementary school to postgraduate degree. Opinion questions may be asked in yes-or-no terms, or on a five- or seven-point scale with a neutral value in the middle. Meanwhile, many numeric measures, such as number of children, may still take only integer values within a relatively narrow range. In practice these too may be treated as ordered categorical variables running from zero to some top-coded value such as "Six or more." Even properly continuous measures, such as income, are rarely reported to the dollar and are often obtainable only as ordered categories. The GSS data in `gss_sm` contains many measures of this sort. You can take a peek at it, as usual, by typing its name at the console. You could also try `glimpse(gss_sm)`, which will give a compact summary of all the variables in the data.

We will make a smoothed scatterplot (fig. 4.5) of the relationship between the age of the respondent and the number of children they have. In `gss_sm` the `childs` variable is a numeric count of the respondent's children. (There is also a variable named `kids` that

To begin with, we will use the GSS data in a slightly naive way. In particular we will not consider sample weights when making the figures in this chapter. In chapter 6 we will learn how to calculate frequencies and other statistics from data with a complex or weighted survey design.

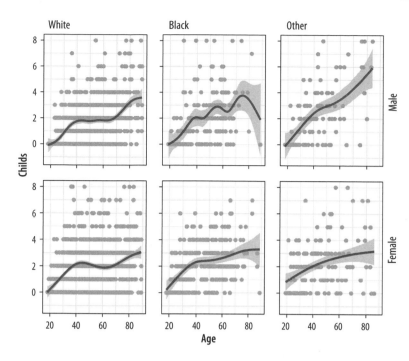

Figure 4.5: Faceting on two categorical variables. Each panel plots the relationship between age and number of children, with the facets breaking out the data by sex (in the rows) and race (in the columns).

is the same measure, but its class is an ordered factor rather than a number.) We will then facet this relationship by sex and race of the respondent. We use R's formula notation in the facet_grid function to facet sex and race. This time, because we are cross-classifying our results, the formula is two-sided: facet_grid(sex ~ race).

```
p ← ggplot(data = gss_sm,
          mapping = aes(x = age, y = childs))
p + geom_point(alpha = 0.2) +
    geom_smooth() +
    facet_grid(sex ~ race)
```

Multipanel layouts of this kind are especially effective when used to summarize continuous variation (as in a scatterplot) across two or more categorical variables, with the categories (and hence the panels) ordered in some sensible way. We are not limited to two-way comparison. Further categorical variables can be added to the formula, too, (e.g., sex ~ race + degree) for more complex multiway plots. However, the multiple dimensions of plots like this will quickly become very complicated if the variables have more than a few categories each.

4.4 Geoms Can Transform Data

We have already seen several examples where geom_smooth() was included as a way to add a trend line to the figure. Sometimes we plotted a LOESS line, sometimes a straight line from an OLS regression, and sometimes the result of a Generalized Additive Model. We did not have to have any strong idea of the differences between these methods. Neither did we have to write any code to specify the underlying models, beyond telling the method argument in geom_smooth() which one we wanted to use. The geom_smooth() function did the rest.

Thus some geoms plot our data directly on the figure, as is the case with geom_point(), which takes variables designated as x and y and plots the points on a grid. But other geoms clearly do more work on the data before it gets plotted. Every geom_ function has an associated stat_ function that it uses by default. The reverse is also the case: every stat_ function has an associated geom_ function that it will plot by default if you ask it to. This is not particularly important to know by itself, but as we will see in the next section, we sometimes want to calculate a different statistic for the geom from the default.

Sometimes the calculations being done by the stat_ functions that work together with the geom_ functions might not be immediately obvious. For example, consider figure 4.6, produced by a new geom, geom_bar().

Try p + stat_smooth(), for example.

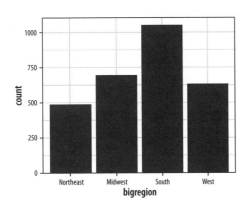

Figure 4.6: A bar chart.

```
p ← ggplot(data = gss_sm, mapping = aes(x = bigregion))
p + geom_bar()
```

Here we specified just one mapping, aes(x = bigregion). The bar chart produced gives us a count of the number of (individual) observations in the data set by region of the United States. This seems sensible. But there is a y-axis variable here, count, that is not in the data. It has been calculated for us. Behind the scenes, geom_bar called the default stat_ function associated with it, stat_count(). This function computes two new variables, count and prop (short for proportion). The count statistic is the one geom_bar() uses by default.

```
p ← ggplot(data = gss_sm, mapping = aes(x = bigregion))
p + geom_bar(mapping = aes(y = ..prop..))
```

If we want a chart of relative frequencies rather than counts, we will need to get the prop statistic instead. When ggplot calculates the count or the proportion, it returns temporary variables that we can use as mappings in our plots. The relevant statistic is called ..prop.. rather than prop. To make sure these temporary variables won't be confused with others we are working with, their names begin and end with two periods. (This is because we might already have a variable called count or prop in our dataset.) So our calls to it from the aes() function will generically look like this: <mapping> = <..statistic..>. In this case, we want y to use the calculated proportion, so we say aes(y = ..prop..).

The resulting plot in figure 4.7 is still not right. We no longer have a count on the y-axis, but the proportions of the bars all have a value of 1, so all the bars are the same height. We want them to *sum* to 1, so that we get the number of observations per continent as a proportion of the total number of observations, as in figure 4.8. This is a grouping issue again. In a sense, it's the reverse of the earlier grouping problem we faced when we needed to tell ggplot that our yearly data was grouped by country. In this case, we need to tell ggplot to *ignore* the x-categories when calculating denominator of the proportion and use the total number observations instead. To do so we specify group = 1 inside the aes() call. The value of 1 is just a kind of "dummy group" that tells ggplot to use the whole dataset when establishing the denominator for its prop calculations.

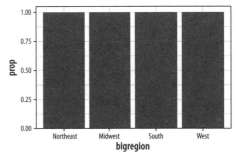

Figure 4.7: A first go at a bar chart with proportions.

```
p ← ggplot(data = gss_sm, mapping = aes(x = bigregion))
p + geom_bar(mapping = aes(y = ..prop.., group = 1))
```

Let's look at another question from the survey. The gss_sm data contains a religion variable derived from a question asking "What is your religious preference? Is it Protestant, Catholic, Jewish, some other religion, or no religion?"

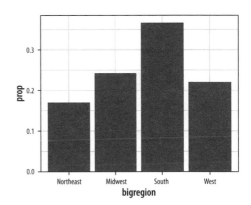

Figure 4.8: A bar chart with correct proportions.

```
table(gss_sm$religion)
```

```
##
## Protestant   Catholic     Jewish       None      Other
##       1371        649         51        619        159
```

To graph this, we want a bar chart with religion on the x axis (as a categorical variable), and with the bars in the chart also

Recall that the $ character is one way of accessing individual columns within a data frame or tibble.

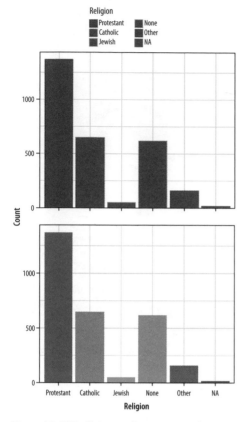

Figure 4.9: GSS religious preference mapped to color (*top*) and both color and fill (*bottom*).

colored by religion. If the gray bars look boring and we want to fill them with color instead, we can map the religion variable to fill in addition to mapping it to x. Remember, fill is for painting the insides of shapes. If we map religion to color, only the border lines of the bars will be assigned colors, and the insides will remain gray.

```
p ← ggplot(data = gss_sm, mapping = aes(x = religion, color = religion))
p + geom_bar()

p ← ggplot(data = gss_sm, mapping = aes(x = religion, fill = religion))
p + geom_bar() + guides(fill = FALSE)
```

By doing this, we have mapped two aesthetics to the same variable. Both x and fill are mapped to religion. There is nothing wrong with this. However, these are still two separate mappings, and so they get two separate scales (fig. 4.9). The default is to show a legend for the color variable. This legend is redundant because the categories of religion are already separated out on the x-axis. In its simplest use, the guides() function controls whether guiding information about any particular mapping appears or not. If we set guides(fill = FALSE), the legend is removed, in effect saying that the viewer of the figure does not need to be shown any guiding information about this mapping. Setting the guide for some mapping to FALSE has an effect only if there is a legend to turn off to begin with. Trying x = FALSE or y = FALSE will have no effect, as these mappings have no additional guides or legends separate from their scales. It is possible to turn the x and y scales off altogether, but this is done through a different function, one from the scale_ family.

4.5 Frequency Plots the Slightly Awkward Way

A more appropriate use of the fill aesthetic with geom_bar() is to cross-classify two categorical variables. This is the graphical equivalent of a frequency table of counts or proportions. Using the GSS data, for instance, we might want to examine the distribution of religious preferences within different regions of the United States. In the next few paragraphs we will see how to do this just using ggplot. However, as we shall also discover, it is often not the most transparent way to make frequency tables of this sort. The next chapter introduces a simpler and less error-prone approach where we calculate the table first before passing the results along to ggplot to graph. As you work through this section, bear in mind that if you find things slightly awkward or confusing it is because that's exactly what they are.

Let's say we want to look at religious preference by census region. That is, we want the `religion` variable broken down proportionally within `bigregion`. When we cross-classify categories in bar charts, there are several ways to display the results. With `geom_bar()` the output is controlled by the `position` argument. Let's begin by mapping fill to `religion`.

```
p ← ggplot(data = gss_sm, mapping = aes(x = bigregion, fill = religion))
p + geom_bar()
```

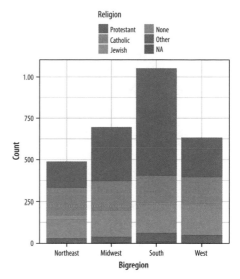

Figure 4.10: A stacked bar chart of religious preference by census region.

The default output of `geom_bar()` is a stacked bar chart (fig. 4.10) with counts on the y-axis (and hence counts within the stacked segments of the bars also). Region of the country is on the x-axis, and counts of religious preference are stacked within the bars. As we saw in chapter 1, it is somewhat difficult for readers of the chart to compare lengths and areas on an unaligned scale. So while the relative position of the bottom categories are quite clear (thanks to them all being aligned on the x-axis), the relative positions of, say, the "Catholic" category is harder to assess. An alternative choice is to set the `position` argument to `"fill."` (This is different from the fill aesthetic.)

```
p ← ggplot(data = gss_sm, mapping = aes(x = bigregion, fill = religion))
p + geom_bar(position = "fill")
```

Now (fig. 4.11) the bars are all the same height, which makes it easier to compare proportions across groups. But we lose the ability to see the relative size of each cut with respect to the overall total. What if we wanted to show the proportion or percentage of religions within regions of the country, like in figure 4.11, but instead of stacking the bars we wanted separate bars instead? As a first attempt, we can use `position = "dodge"` to make the bars within each region of the country appear side by side. However, if we do it this way (try it), we will find that ggplot places the bars side-by-side as intended but changes the y-axis back to a *count* of

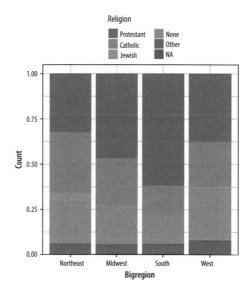

Figure 4.11: Using the fill position adjustment to show relative proportions across categories.

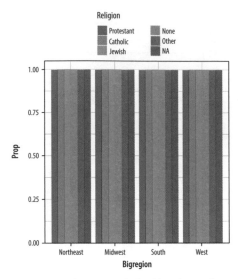

Figure 4.12: A first go at a dodged bar chart with proportional bars.

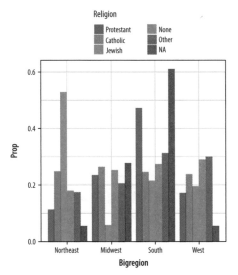

Figure 4.13: A second attempt at a dodged bar chart with proportional bars.

Proportions for smaller subpopulations tend to bounce around from year to year in the GSS.

cases within each category rather than showing us a proportion. We saw in figure 4.8 that to display a proportion we needed to map `y = ..prop..`, so the correct statistic would be calculated. Let's see if that works.

```
p ← ggplot(data = gss_sm, mapping = aes(x = bigregion, fill = religion))
p + geom_bar(position = "dodge", mapping = aes(y = ..prop..))
```

The result (fig. 4.12) is certainly colorful but not what we wanted. Just as in figure 4.7, there seems to be an issue with the grouping. When we just wanted the overall proportions for one variable, we mapped `group = 1` to tell ggplot to calculate the proportions with respect to the overall N. In this case our grouping variable is `religion`, so we might try mapping that to the `group` aesthetic.

```
p ← ggplot(data = gss_sm, mapping = aes(x = bigregion, fill = religion))
p + geom_bar(position = "dodge", mapping = aes(y = ..prop..,
    group = religion))
```

This gives us a bar chart where the values of `religion` are broken down across regions, with a proportion showing on the y-axis. If you inspect the bars in figure 4.13, you will see that they do not sum to one within each region. Instead, the bars for any particular religion sum to one *across* regions.

This lets us see that nearly half of those who said they were Protestant live in the South, for example. Meanwhile, just over 10 percent of those saying they were Protestant live in the Northeast. Similarly, it shows that over half of those saying they were Jewish live in the Northeast, compared to about a quarter who live in the South.

We are still not quite where we originally wanted to be. Our goal was to take the stacked bar chart in Figure 4.10 but have the proportions shown side-by-side instead of on top of one another.

```
p ← ggplot(data = gss_sm, mapping = aes(x = religion))
p + geom_bar(position = "dodge", mapping = aes(y = ..prop..,
    group = bigregion)) + facet_wrap(~bigregion, ncol = 1)
```

It turns out that the easiest thing to do is to stop trying to force `geom_bar()` to do all the work in a single step. Instead, we can ask

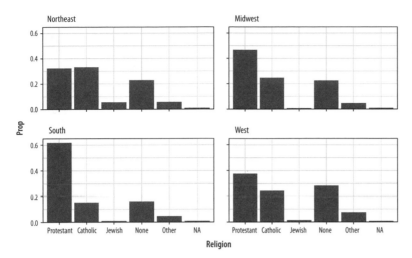

Figure 4.14: Faceting proportions within region.

ggplot to give us a proportional bar chart of religious affiliation, and then facet that by region. The proportions in figure 4.14 are calculated within each panel, which is the breakdown we wanted. This has the added advantage of not producing too many bars within each category.

We could polish this plot further, but for the moment we will stop here. When constructing frequency plots directly in ggplot, it is easy to get stuck in a cycle of not quite getting the marginal comparison that you want, and more or less randomly poking at the mappings to try to stumble on the right breakdown. In the next chapter, we will learn how to use the tidyverse's dplyr library to produce the tables we want *before* we try to plot them. This is a more reliable approach, and easier to check for errors. It will also give us tools that can be used for many more tasks than producing summaries.

4.6 Histograms and Density Plots

Different geoms transform data in different ways, but ggplot's vocabulary for them is consistent. We can see similar transformations at work when summarizing a continuous variable using a histogram, for example. A histogram is a way of summarizing a continuous variable by chopping it up into segments or "bins" and counting how many observations are found within each bin. In a bar chart, the categories are given to us going in (e.g., regions of the country, or religious affiliation). With a histogram, we have to decide how finely to bin the data.

For example, ggplot comes with a dataset, midwest, containing information on counties in several midwestern states of the United States. Counties vary in size, so we can make a histogram showing the distribution of their geographical areas. Area is measured in square miles. Because we are summarizing a continuous variable using a series of bars, we need to divide the observations into groups, or bins, and count how many are in each one. By default, the geom_histogram() function will choose a bin size for us based on a rule of thumb.

```
p ← ggplot(data = midwest, mapping = aes(x = area))
p + geom_histogram()
```

```
## `stat_bin()` using `bins = 30`. Pick better value with
## `binwidth`.
```

```
p ← ggplot(data = midwest, mapping = aes(x = area))
p + geom_histogram(bins = 10)
```

As with the bar charts, a newly calculated variable, count, appears on the x-axis in figure 4.15. The notification from R tells us that behind the scenes the stat_bin() function picked thirty bins, but we might want to try something else. When drawing histograms it is worth experimenting with bins and also optionally the origin of the x-axis. Each, especially bins, will make a big difference in how the resulting figure looks.

While histograms summarize single variables, it's also possible to use several at once to compare distributions. We can facet histograms by some variable of interest, or as here, we can compare them in the same plot using the fill mapping (fig. 4.16).

Figure 4.15: Histograms of the same variable, using different numbers of bins.

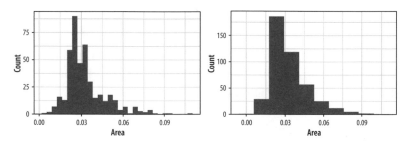

```
oh_wi ← c("OH", "WI")

p ← ggplot(data = subset(midwest, subset = state %in% oh_wi),
    mapping = aes(x = percollege, fill = state))
p + geom_histogram(alpha = 0.4, bins = 20)
```

We subset the data here to pick out just two states. To do this we create a character vector with just two elements, "OH" and "WI." Then we use the subset() function to take our data and filter it so that we select only rows whose state name is in this vector. The %in% operator is a convenient way to filter on more than one term in a variable when using subset().

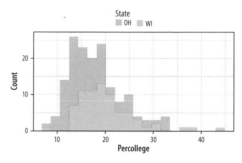

Figure 4.16: Comparing two histograms.

When working with a continuous variable, an alternative to binning the data and making a histogram is to calculate a kernel density estimate of the underlying distribution. The geom_density() function will do this for us (fig. 4.17).

```
p ← ggplot(data = midwest, mapping = aes(x = area))
p + geom_density()
```

We can use color (for the lines) and fill (for the body of the density curve) here, too. These figures often look quite nice. But when there are several filled areas on the plot, as in this case, the overlap can become hard to read. (Figures 4.18 and 4.19 are examples.) If you want to make the baselines of the density curves go away, you can use geom_line(stat = "density") instead. This also removes the possibility of using the fill aesthetic. But this may be an improvement in some cases. Try it with the plot of state areas and see how they compare.

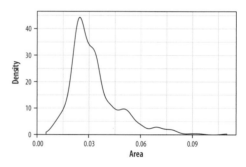

Figure 4.17: Kernel density estimate of county areas.

```
p ← ggplot(data = midwest, mapping = aes(x = area, fill = state,
    color = state))
p + geom_density(alpha = 0.3)
```

Just like geom_bar(), the count-based defaults computed by the stat_ functions used by geom_histogram() and geom_density() will return proportional measures if we ask them. For geom_density(), the stat_density() function can return its default ..density.. statistic, or ..scaled.., which will give a proportional density estimate. It can also return a statistic called

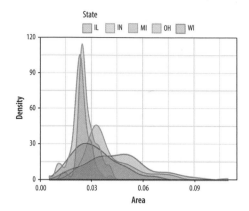

Figure 4.18: Comparing distributions.

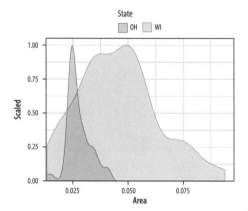

Figure 4.19: Scaled densities.

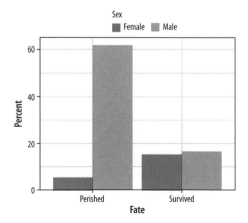

Figure 4.20: Survival on the *Titanic*, by sex.

..count.., which is the density times the number of points. This can be used in stacked density plots.

```
p ← ggplot(data = subset(midwest, subset = state %in% oh_wi),
    mapping = aes(x = area, fill = state, color = state))
p + geom_density(alpha = 0.3, mapping = (aes(y = ..scaled..)))
```

4.7 Avoid Transformations When Necessary

As we have seen from the beginning, ggplot normally makes its charts starting from a full dataset. When we call geom_bar() it does its calculations on the fly using stat_count() behind the scenes to produce the counts or proportions it displays. In the previous section, we looked at a case where we wanted to group and aggregate our data ourselves before handing it off to ggplot. But often our data is, in effect, *already* a summary table. This can happen when we have computed a table of marginal frequencies or percentages from the original data. Plotting results from statistical models also puts us in this position, as we will see later. Or it may be that we just have a finished table of data (from the census, say, or an official report) that we want to make into a graph. For example, perhaps we do not have the individual-level data on who survived the *Titanic* disaster, but we do have a small table of counts of survivors by sex:

```
titanic
```

```
##       fate    sex    n percent
## 1 perished   male 1364    62.0
## 2 perished female  126     5.7
## 3 survived   male  367    16.7
## 4 survived female  344    15.6
```

Because we are working directly with percentage values in a summary table, we no longer have any need for ggplot to count up values for us or perform any other calculations. That is, we do not need the services of any stat_ functions that geom_bar() would normally call. We can tell geom_bar() not to do any work on the variable before plotting it. To do this we say stat = 'identity' in the geom_bar() call. We'll also move the legend to the top of the chart (fig. 4.20).

```
p ← ggplot(data = titanic, mapping = aes(x = fate, y = percent,
    fill = sex))
p + geom_bar(position = "dodge", stat = "identity") + theme(legend.position = "top")
```

For convenience ggplot also provides a related geom, geom_col(), which has exactly the same effect but assumes that stat = "identity." We will use this form in the future when we don't need any calculations done on the plot.

The position argument in geom_bar() and geom_col() can also take the value of "identity." Just as stat = "identity" means "don't do any summary calculations," position = "identity" means "just plot the values as given." This allows us to do things like plotting a flow of positive and negative values in a bar chart. This sort of graph is an alternative to a lineplot and is often seen in public policy settings where changes relative to some threshold level or baseline are of interest. For example, the oecd_sum table in socviz contains information on average life expectancy at birth within the United States and across other OECD countries.

```
oecd_sum
```

```
## # A tibble: 57 x 5
## # Groups:   year [57]
##     year other   usa  diff hi_lo
##    <int> <dbl> <dbl> <dbl> <chr>
##  1  1960  68.6  69.9 1.30  Below
##  2  1961  69.2  70.4 1.20  Below
##  3  1962  68.9  70.2 1.30  Below
##  4  1963  69.1  70.0 0.900 Below
##  5  1964  69.5  70.3 0.800 Below
##  6  1965  69.6  70.3 0.700 Below
##  7  1966  69.9  70.3 0.400 Below
##  8  1967  70.1  70.7 0.600 Below
##  9  1968  70.1  70.4 0.300 Below
## 10  1969  70.1  70.6 0.500 Below
## # ... with 47 more rows
```

The other column is the average life expectancy in a given year for OECD countries, excluding the United States. The usa column is the U.S. life expectancy, diff is the difference between the two

The US life expectancy gap

Difference between US and OECD average life expectancies, 1960–2015

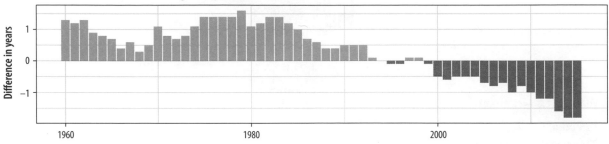

Data: OECD. After a chart by Christopher Ingraham,
Washington Post, December 27th 2017.

Figure 4.21: Using `geom_col()` to plot negative and positive values in a bar chart.

values, and `hi_lo` indicates whether the U.S. value for that year was above or below the OECD average. We will plot the difference over time and use the `hi_lo` variable to color the columns in the chart (fig. 4.21).

```
p ← ggplot(data = oecd_sum,
            mapping = aes(x = year, y = diff, fill = hi_lo))
p + geom_col() + guides(fill = FALSE) +
  labs(x = NULL, y = "Difference in Years",
       title = "The US Life Expectancy Gap",
       subtitle = "Difference between US and OECD
                    average life expectancies, 1960-2015",
       caption = "Data: OECD. After a chart by Christopher Ingraham,
                    Washington Post, December 27th 2017.")
```

As with the `titanic` plot, the default action of `geom_col()` is to set both `stat` and `position` to "identity." To get the same effect with `geom_bar()` we would need to say `geom_bar(position = "identity")`. As before, the `guides(fill = FALSE)` instruction at the end tells ggplot to drop the unnecessary legend that would otherwise be automatically generated to accompany the fill mapping.

At this point, we have a pretty good sense of the core steps we must take to visualize our data. In fact, thanks to ggplot's default settings, we now have the ability to make good-looking and informative plots. Starting with a tidy dataset, we know how to map variables to aesthetics, choose from a variety of geoms, and make some adjustments to the scales of the plot. We also know more about selecting the right sort of computed statistic to show on the graph, if that's what's needed, and how to facet our core plot by one or more variables. We know how to set descriptive labels for axes and write a title, subtitle, and caption. Now we're in a position to put these skills to work in a more fluent way.

4.8 Where to Go Next

- Revisit the `gapminder` plots at the beginning of the chapter and experiment with different ways to facet the data. Try plotting population and per capita GDP while faceting on year, or even on country. In the latter case you will get a lot of panels, and plotting them straight to the screen may take a long time. Instead, assign the plot to an object and save it as a PDF file to your `figures/` folder. Experiment with the height and width of the figure.
- Investigate the difference between a formula written as `facet_grid(sex ~ race)` and one written as `facet_grid(~ sex + race)`.
- Experiment to see what happens when you use `facet_wrap()` with more complex formulas like `facet_wrap(~ sex + race)` instead of `facet_grid`. Like `facet_grid()`, the `facet_wrap()` function can facet on two or more variables at once. But it will do it by laying the results out in a wrapped one-dimensional table instead of a fully cross-classified grid.
- Frequency polygons are closely related to histograms. Instead of displaying the count of observations using bars, they display it with a series of connected lines. You can try the various `geom_histogram()` calls in this chapter using `geom_freqpoly()` instead.
- A histogram bins observations for one variable and shows a bar with the count in each bin. We can do this for two variables at once, too. The `geom_bin2d()` function takes two mappings, `x` and `y`. It divides your plot into a grid and colors the bins by the count of observations in them. Try using it on the `gapminder` data to plot life expectancy versus per capita GDP. Like a histogram, you can vary the number or width of the bins for both `x` or `y`. Instead of saying `bins = 30` or `binwidth = 1`, provide a number for both `x` and `y` with, for example, `bins = c(20, 50)`. If you specify `binwidth` instead, you will need to

pick values that are on the same scale as the variable you are mapping.

- Density estimates can also be drawn in two dimensions. The `geom_density_2d()` function draws contour lines estimating the joint distribution of two variables. Try it with the `midwest` data, for example, plotting percent below the poverty line (`percbelowpoverty`) against percent college-educated (`percollege`). Try it with and without a `geom_point()` layer.

5 Graph Tables, Add Labels, Make Notes

This chapter builds on the foundation we have laid down. Things will get a little more sophisticated in three ways. First, we will learn about how to transform data *before* we send it to ggplot to be turned into a figure. As we saw in chapter 4, ggplot's geoms will often summarize data for us. While convenient, this can sometimes be awkward or even a little opaque. Often it's better to get things into the right shape before we send anything to ggplot. This is a job for another tidyverse component, the `dplyr` library. We will learn how to use some of its "action verbs" to select, group, summarize, and transform our data.

Second, we will expand the number of geoms we know about and learn more about how to choose between them. The more we learn about ggplot's geoms, the easier it will be to pick the right one given the data we have and the visualization we want. As we learn about new geoms, we will also get more adventurous and depart from some of ggplot's default arguments and settings. We will learn how to reorder the variables displayed in our figures, and how to subset the data we use before we display it.

Third, this process of gradual customization will give us the opportunity to learn more about the scale, guide, and theme functions that we have mostly taken for granted until now. These will give us even more control over the content and appearance of our graphs. Together, these functions can be used to make plots much more legible to readers. They allow us to present our data in a more structured and easily comprehensible way, and to pick out the elements of it that are of particular interest. We will begin to use these methods to layer geoms on top of one another, a technique that will allow us to produce sophisticated graphs in a systematic, comprehensible way.

Our basic approach will not change. No matter how complex our plots get, or how many individual steps we take to layer and tweak their features, underneath we will always be doing the same thing. We want a table of tidy data, a mapping of variables to aesthetic elements, and a particular type of graph. If you can keep sight of this, it will make it easier to confidently approach the job of getting any particular graph to look just right.

5.1 Use Pipes to Summarize Data

In chapter 4 we began making plots of the distributions and relative frequencies of variables. Cross-classifying one measure by another is one of the basic descriptive tasks in data analysis. Tables 5.1 and 5.2 show two common ways of summarizing our GSS data on the distribution of religious affiliation and region. Table 5.1 shows the column marginals, where the numbers sum to a hundred by column and show, e.g., the distribution of Protestants across regions. Meanwhile in table 5.2 the numbers sum to a hundred across the rows, showing, for example, the distribution of religious affiliations within any particular region.

We saw in chapter 4 that geom_bar() can plot both counts and relative frequencies depending on what we asked of it. In practice, though, letting the geoms (and their stat_ functions) do the work can sometimes get a little confusing. It is too easy to lose track of whether one has calculated row margins, column margins, or

TABLE 5.1
Column marginals. (Numbers in columns sum to 100.)

	Protestant	Catholic	Jewish	None	Other	NA
Northeast	12	25	53	18	18	6
Midwest	24	27	6	25	21	28
South	47	25	22	27	31	61
West	17	24	20	29	30	6

TABLE 5.2
Row marginals. (Numbers in rows sum to 100.)

	Protestant	Catholic	Jewish	None	Other	NA
Northeast	32	33	6	23	6	0
Midwest	47	25	0	23	5	1
South	62	15	1	16	5	1
West	38	25	2	28	8	0

overall relative frequencies. The code to do the calculations on the fly ends up stuffed into the mapping function and can become hard to read. A better strategy is to calculate the frequency table you want first and then plot that table. This has the benefit of allowing you do to some quick sanity checks on your tables, to make sure you haven't made any errors.

Let's say we want a plot of the row marginals for religion within region. We will take the opportunity to do a little bit of data munging in order to get from our underlying table of GSS data to the summary tabulation that we want to plot. To do this we will use the tools provided by dplyr, a component of the tidyverse that provides functions for manipulating and reshaping tables of data on the fly. We start from our individual-level gss_sm data frame with its bigregion and religion variables. Our goal is a summary table with percentages of religious preferences grouped within region.

As shown schematically in figure 5.1, we will start with our individual-level table of about 2,500 GSS respondents. Then we want to summarize them into a new table that shows a count of each religious preference, grouped by region. Finally we will turn these within-region counts into percentages, where the denominator is the total number of respondents within each region. The dplyr library provides a few tools to make this easy and clear to read. We will use a special operator, %>%, to do our work. This is the *pipe* operator. It plays the role of the yellow triangle in figure 5.1, in that it helps us perform the actions that get us from one table to the next.

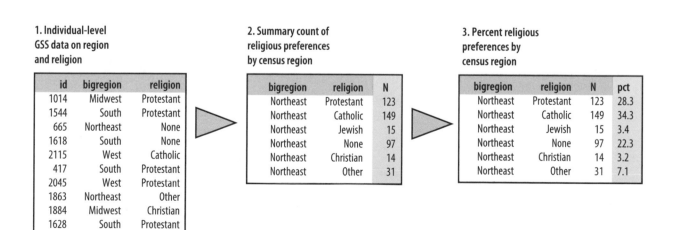

Figure 5.1: How we want to transform the individual-level data.

We have been building our plots in an *additive* fashion, starting with a ggplot object and layering on new elements. By analogy, think of the %>% operator as allowing us to start with a data frame and perform a *sequence* or *pipeline* of operations to turn it into another, usually smaller and more aggregated, table. Data goes in one side of the pipe, actions are performed via functions, and results come out the other. A pipeline is typically a series of operations that do one or more of four things:

group_by()

- *Group* the data into the nested structure we want for our summary, such as "Religion by Region" or "Authors by Publications by Year."

filter() rows; select() columns

- *Filter* or *select* pieces of the data by row, column, or both. This gets us the piece of the table we want to work on.

mutate()

- *Mutate* the data by creating new variables at the *current* level of grouping. This adds new columns to the table without aggregating it.

summarize()

- *Summarize* or aggregate the grouped data. This creates new variables at a *higher* level of grouping. For example we might calculate means with mean() or counts with n(). This results in a smaller, summary table, which we might further summarize or mutate if we want.

We use the dplyr functions group_by(), filter(), select(), mutate(), and summarize() to carry out these tasks within our pipeline. They are written in a way that allows them to be easily piped. That is, they understand how to take inputs from the left side of a pipe operator and pass results along through the right side of one. The dplyr documentation has some useful vignettes that introduce these grouping, filtering, selection, and transformation functions. There is also a more detailed discussion of these tools, along with many more examples, in Wickham & Grolemund (2016).

We will create a new table called rel_by_region. Here's the code:

```
rel_by_region ← gss_sm %>%
    group_by(bigregion, religion) %>%
    summarize(N = n()) %>%
    mutate(freq = N / sum(N),
           pct = round((freq*100), 0))
```

What are these lines doing? First, we are creating an object as usual, with the familiar assignment operator, <-. Next comes the pipeline. Read the objects and functions from left to right, with the pipe operator "%>%" connecting them together meaning "and then...." Objects on the left side "pass through" the pipe, and whatever is specified on the right of the pipe gets done to that object. The resulting object then passes through to the right again, and so on down to the end of the pipeline.

Reading from the left, the code says this:

- Create a new object, rel_by_region. It will get the result of the following sequence of actions: Start with the gss_sm data, and then

`rel_by_region ← gss_sm %>%`

- Group the rows by bigregion and, within that, by religion.

`group_by(bigregion, religion) %>%`

- Summarize this table to create a new, much smaller table, with three columns: bigregion, religion, and a new summary variable, N, that is a count of the number of observations within each religious group for each region.

`summarize(N = n()) %>%`

- With this new table, use the N variable to calculate two new columns: the relative proportion (freq) and percentage (pct) for each religious category, still grouped by region. Round the results to the nearest percentage point.

`mutate(freq = N / sum(N), pct = round((freq*100), 0))`

In this way of doing things, objects passed along the pipeline and the functions acting on them carry some assumptions about their context. For one thing, you don't have to keep specifying the name of the underlying data frame object you are working from. Everything is implicitly carried forward from gss_sm. Within the pipeline, the transient or implicit objects created from your summaries and other transformations are carried through, too.

Second, the group_by() function sets up how the grouped or nested data will be processed within the summarize() step. Any function used to create a new variable within summarize(), such as mean() or sd() or n(), will be applied to the *innermost* grouping level first. Grouping levels are named from left to right within group_by() from outermost to innermost. So the function call summarize(N = n()) counts up the number of observations for each value of religion within bigregion and puts them in a new variable named N. As dplyr's functions see things, summarizing actions peel off one grouping level at a time, so that the resulting summaries are at the next level up. In this case, we start with individual-level observations and group them by religion within region. The summarize() operation aggregates the individual observations to counts of the number of people affiliated with each religion, for each region.

Third, the mutate() step takes the N variable and uses it to create freq, the relative frequency for each subgroup within region, and finally pct, the relative frequency turned into a rounded percentage. These mutate() operations add or remove columns from tables but do not change the grouping level.

Inside both mutate() and summarize(), we are able to create new variables in a way that we have not seen before. Usually, when we see something like name = value inside a function, the name is a general, named argument and the function is expecting information from us about the specific value it should take. Normally if we give a function a named argument it doesn't know about (aes(chuckles = year)), it will ignore it, complain, or break. With summarize() and mutate(), however, we can invent named arguments. We are still assigning specific values to N, freq, and pct, but we pick the names, too. They are the names that the newly created variables in the summary table will have. The summarize() and mutate() functions do not need to know what they will be in advance.

Finally, when we use mutate() to create the freq variable, not only can we make up that name within the function, mutate() is also clever enough to let us *use* that name right away, on the next line of the same function call, when we create the pct variable. This means we do not have to repeatedly write separate mutate() calls for every new variable we want to create.

Our pipeline takes the gss_sm data frame, which has 2,867 rows and 32 columns, and transforms it into rel_by_region, a summary table with 24 rows and 5 columns that looks like this, in part:

As in the case of aes(x = gdpPercap, y = lifeExp), for example.

rel_by_region

```
## # A tibble: 24 x 5
## # Groups:   bigregion [4]
##    bigregion religion       N    freq  pct
##    <fct>     <fct>      <int>   <dbl> <dbl>
##  1 Northeast Protestant   158 0.324    32.
##  2 Northeast Catholic     162 0.332    33.
```

```
## 3 Northeast Jewish        27 0.0553      6.
## 4 Northeast None         112 0.230      23.
## 5 Northeast Other         28 0.0574      6.
## 6 Northeast <NA>           1 0.00205     0.
## 7 Midwest   Protestant   325 0.468      47.
## 8 Midwest   Catholic     172 0.247      25.
## 9 Midwest   Jewish         3 0.00432     0.
## 10 Midwest  None         157 0.226      23.
## # ... with 14 more rows
```

The variables specified in `group_by()` are retained in the new summary table; the variables created with `summarize()` and `mutate()` are added, and all the other variables in the original dataset are dropped.

We said before that, when trying to grasp what each additive step in a `ggplot()` sequence does, it can be helpful to work backward, removing one piece at a time to see what the plot looks like when that step is not included. In the same way, when looking at pipelined code it can be helpful to start from the end of the line and then remove one "%>%" step at a time to see what the resulting intermediate object looks like. For instance, what if we remove the `mutate()` step from the code above? What does `rel_by_region` look like then? What if we remove the `summarize()` step? How big is the table returned at each step? What level of grouping is it at? What variables have been added or removed?

Plots that do not require sequential aggregation and transformation of the data before they are displayed are usually easy to write directly in ggplot, as the details of the layout are handled by a combination of mapping variables and layering geoms. One-step filtering or aggregation of the data (such as calculating a proportion, or a specific subset of observations) is also straightforward. But when the result we want to display is several steps removed from the data, and in particular when we want to group or aggregate a table and do some more calculations on the result before drawing anything, then it can make sense to use dplyr's tools to produce these summary tables first. This is true even if it would be possible to do it within a `ggplot()` call. In addition to making our code easier to read, pipes let us more easily perform sanity checks on our results, so that we are sure we have grouped and summarized things in the right order. For instance, if we have done things properly with rel_by_region, the pct values associated with religion should sum to 100 within each region, perhaps with a bit of rounding error. We can quickly check this using a very short pipeline:

```
rel_by_region %>% group_by(bigregion) %>% summarize(total = sum(pct))
```

```
## # A tibble: 4 x 2
##   bigregion total
##   <fct>     <dbl>
## 1 Northeast  100.
## 2 Midwest    101.
## 3 South      100.
## 4 West       101.
```

This looks good. As before, now that we are working directly with percentage values in a summary table, we can use geom_col() instead of geom_bar().

```
p <- ggplot(rel_by_region, aes(x = bigregion, y = pct, fill = religion))
p + geom_col(position = "dodge2") +
    labs(x = "Region",y = "Percent", fill = "Religion") +
    theme(legend.position = "top")
```

Try going back to the code for figure 4.13, in chapter 4, and using this "dodge2" argument instead of the "dodge" argument there.

We use a different position argument here, dodge2 instead of dodge. This puts the bars side by side. When dealing with pre-computed values in geom_col(), the default position is to make a proportionally stacked column chart. If you use dodge they will be stacked within columns, but the result will read incorrectly. Using dodge2 puts the subcategories (religious affiliations) side-by-side within groups (regions).

The values in the bar chart in figure 5.2 are the percentage equivalents to the stacked counts in figure 4.10. Religious affiliations sum to 100 percent within region. The trouble is, although we now know how to cleanly produce frequency tables, this is still a bad figure! It is too crowded, with too many bars side by side. We can do better.

As a rule, dodged charts can be more cleanly expressed as faceted plots. Faceting removes the need for a legend and thus makes the chart simpler to read. We also introduce a new function. If we map religion to the x-axis, the labels will overlap and become

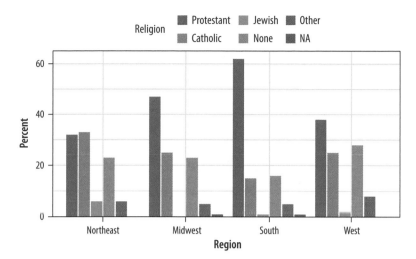

illegible. It's possible to manually adjust the tick-mark labels so that they are printed at an angle, but that isn't so easy to read, either. It makes more sense to put the religions on the y-axis and the percent scores on the x-axis. Because of the way geom_bar() works internally, simply swapping the x and y mapping will not work. (Try it and see what happens.) What we do instead is to transform the *coordinate system* that the results are plotted in, so that the x and y axes are flipped. We do this with coord_flip().

```
p ← ggplot(rel_by_region, aes(x = religion, y = pct, fill = religion))
p + geom_col(position = "dodge2") +
    labs(x = NULL, y = "Percent", fill = "Religion") +
    guides(fill = FALSE) +
    coord_flip() +
    facet_grid(~ bigregion)
```

For most plots the coordinate system is Cartesian, showing plots on a plane defined by an x-axis and a y-axis. The coord_cartesian() function manages this, but we don't need to call it. The coord_flip() function switches the x and y axes after the plot is made. It does not remap variables to aesthetics. In this case, religion is still mapped to x and pct to y. Because the religion names do not need an axis label to be understood, we set x = NULL in the labs() call. (See fig. 5.3.)

We will see more of what dplyr's grouping and filtering operations can do later. It is a flexible and powerful framework. For now, think of it as a way to quickly summarize tables of data

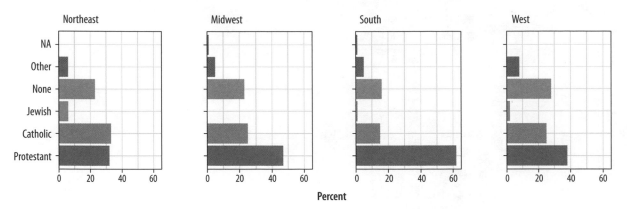

Figure 5.3: Religious preferences by region, faceted version.

without having to write code in the body of our ggplot() or geom_ functions.

5.2 Continuous Variables by Group or Category

Let's move to a new dataset, the organdata table. Like gapminder, it has a country-year structure. It contains a little more than a decade's worth of information on the donation of organs for transplants in seventeen OECD countries. The organ procurement rate is a measure of the number of human organs obtained from cadaver organ donors for use in transplant operations. Along with this donation data, the dataset has a variety of numerical demographic measures, and several categorical measures of health and welfare policy and law. Unlike the gapminder data, some observations are missing. These are designated with a value of NA, R's standard code for missing data. The organdata table is included in the socviz library. Load it up and take a quick look. Instead of using head(), for variety this time we will make a short pipeline to select the first six columns of the dataset and then pick five rows at random using a function called sample_n(). This function takes two main arguments. First we provide the table of data we want to sample from. Because we are using a pipeline, this is implicitly passed down from the beginning of the pipe. Then we supply the number of draws we want to make.

Using numbers this way in select() chooses the numbered columns of the data frame. You can also select variable names directly.

```
organdata %>% select(1:6) %>% sample_n(size = 10)
```

```
## # A tibble: 10 x 6
##    country        year       donors   pop pop_dens   gdp
##    <chr>          <date>      <dbl> <int>    <dbl> <int>
##  1 Switzerland    NA             NA    NA       NA    NA
##  2 Switzerland    1997-01-01   14.3  7089     17.2 27675
##  3 United Kingdom 1997-01-01   13.4 58283     24.0 22442
##  4 Sweden         NA             NA  8559     1.90 18660
##  5 Ireland        2002-01-01   21.0  3932     5.60 32571
##  6 Germany        1998-01-01   13.4 82047     23.0 23283
##  7 Italy          NA             NA 56719     18.8 17430
##  8 Italy          2001-01-01   17.1 57894     19.2 25359
##  9 France         1998-01-01   16.5 58398     10.6 24044
## 10 Spain          1995-01-01   27.0 39223     7.75 15720
```

Lets's start by naively graphing some of the data. We can take a look at a scatterplot of donors vs year (fig. 5.4).

```
p ← ggplot(data = organdata, mapping = aes(x = year, y = donors))
p + geom_point()
```

```
## Warning: Removed 34 rows containing missing values
## (geom_point).
```

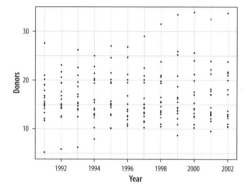

Figure 5.4: Not very informative.

A message from ggplot warns you about the missing values. We'll suppress this warning from now on so that it doesn't clutter the output, but in general it's wise to read and understand the warnings that R gives, even when code appears to run properly. If there are a large number of warnings, R will collect them all and invite you to view them with the warnings() function.

We could use geom_line() to plot each country's time series, like we did with the gapminder data. To do that, remember, we need to tell ggplot what the grouping variable is. This time we can also facet the figure by country (fig. 5.5), as we do not have too many of them.

```
p ← ggplot(data = organdata, mapping = aes(x = year, y = donors))
p + geom_line(aes(group = country)) + facet_wrap(~country)
```

By default the facets are ordered alphabetically by country. We will see how to change this momentarily.

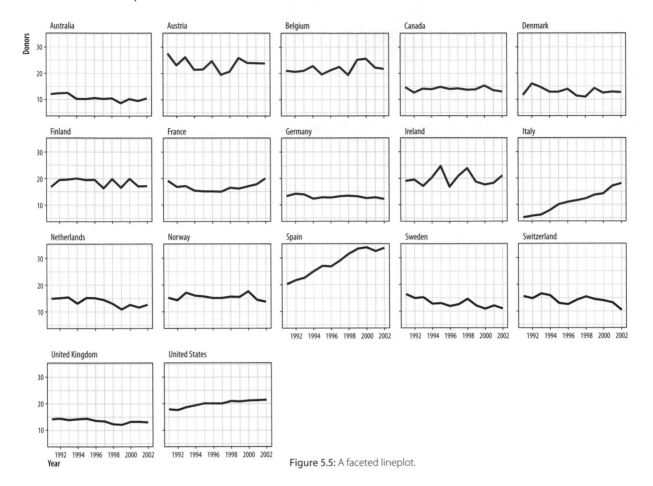

Figure 5.5: A faceted lineplot.

Let's focus on the country-level variation but without paying attention to the time trend. We can use geom_boxplot() to get a picture of variation by year across countries. Just as geom_bar() by default calculates a count of observations by the category you map to x, the stat_boxplot() function that works with geom_boxplot() will calculate a number of statistics that allow the box and whiskers to be drawn. We tell geom_boxplot() the variable we want to categorize by (here, country) and the continuous variable we want summarized (here, donors).

```
p ← ggplot(data = organdata, mapping = aes(x = country, y = donors))
p + geom_boxplot()
```

The boxplots in figure 5.6. look interesting, but two issues could be addressed. First, as we saw in the previous chapter, it

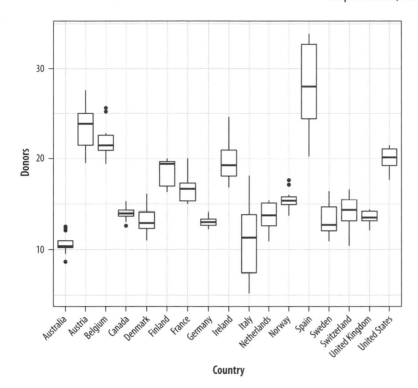

is awkward to have the country names on the x-axis because the labels will overlap. For figure 5.7 use coord_flip() again to switch the axes (but not the mappings).

```
p ← ggplot(data = organdata, mapping = aes(x = country, y = donors))
p + geom_boxplot() + coord_flip()
```

That's more legible but still not ideal. We generally want our plots to present data in some meaningful order. An obvious way is to have the countries listed from high to low average donation rate. We accomplish this by reordering the country variable by the mean of donors. The reorder() function will do this for us. It takes two required arguments. The first is the categorical variable or factor that we want to reorder. In this case, that's country. The second is the variable we want to reorder it by. Here that is the donation rate, donors. The third and optional argument to reorder() is the function you want to use as a summary statistic. If you give reorder() only the first two required arguments, then by default it will reorder the categories of your first variable by the mean value of the second. You can use any sensible function you like to reorder the categorical variable (e.g., median, or sd). There

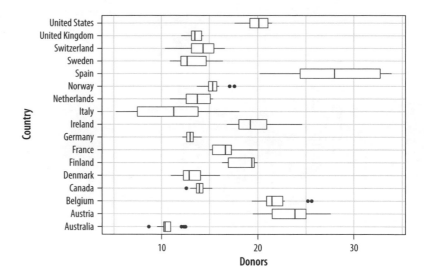

is one additional wrinkle. In R, the default `mean` function will fail with an error if there are missing values in the variable you are trying to take the average of. You must say that it is OK to remove the missing values when calculating the mean. This is done by supplying the `na.rm=TRUE` argument to `reorder()`, which internally passes that argument on to `mean()`. We are reordering the variable we are mapping to the `x` aesthetic, so we use `reorder()` at that point in our code:

```
p ← ggplot(data = organdata, mapping = aes(x = reorder(country,
    donors, na.rm = TRUE), y = donors))
p + geom_boxplot() + labs(x = NULL) + coord_flip()
```

Because it's obvious what the country names are, in the `labs()` call we set their axis label to empty with `labs(x = NULL)`. Ggplot offers some variants on the basic boxplot, including the violin plot. Try redoing figure 5.8 with `geom_violin()`. There are also numerous arguments that control the finer details of the boxes and whiskers, including their width. Boxplots can also take `color` and `fill` aesthetic mappings like other geoms, as in figure 5.9.

```
p ← ggplot(data = organdata,
          mapping = aes(x = reorder(country, donors, na.rm=TRUE),
                        y = donors, fill = world))
p + geom_boxplot() + labs(x=NULL) +
    coord_flip() + theme(legend.position = "top")
```

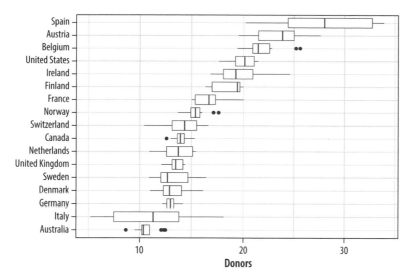

Figure 5.8: Boxplots reordered by mean donation rate.

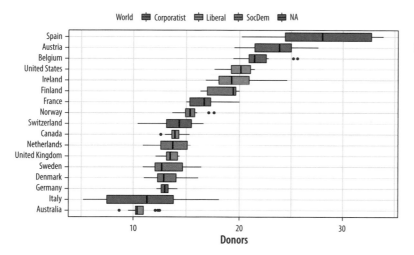

Figure 5.9: A boxplot with the fill aesthetic mapped.

Putting categorical variables on the y-axis to compare their distributions is a useful trick. Its makes it easy to effectively present summary data on more categories. The plots can be quite compact and fit a relatively large number of cases in by row. The approach also has the advantage of putting the variable being compared onto the x-axis, which sometimes makes it easier to compare across categories. If the number of observations within each category is relatively small, we can skip (or supplement) the boxplots and show the individual observations, too. In figure 5.10 we map the world variable to color instead of fill as the default geom_point() plot shape has a color attribute but not a fill.

Figure 5.10: Using points instead of a boxplot.

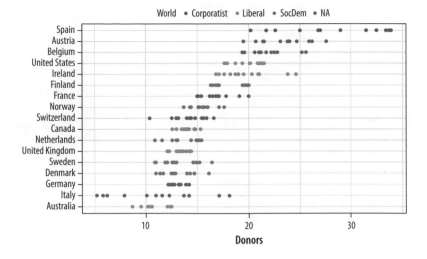

Figure 5.11: Jittering the points.

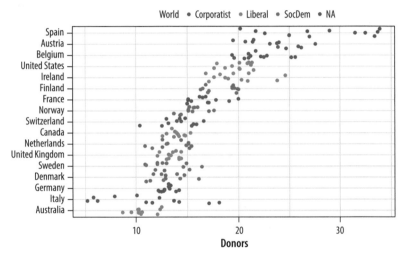

```
p ← ggplot(data = organdata,
           mapping = aes(x = reorder(country, donors, na.rm=TRUE),
                         y = donors, color = world))
p + geom_point() + labs(x=NULL) +
   coord_flip() + theme(legend.position = "top")
```

When we use geom_point() like this, there is some overplotting of observations. In these cases, it can be useful to perturb the data a bit in order to get a better sense of how many observations there are at different values. We use geom_jitter() to do this (fig. 5.11). This geom works much like geom_point() but randomly nudges each observation by a small amount.

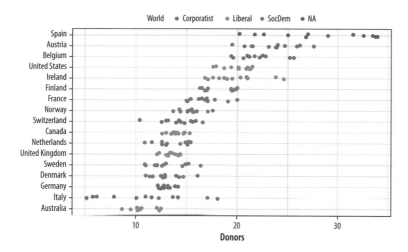

Figure 5.12: A jittered plot.

```
p ← ggplot(data = organdata,
           mapping = aes(x = reorder(country, donors, na.rm=TRUE),
                         y = donors, color = world))
p + geom_jitter() + labs(x=NULL) +
   coord_flip() + theme(legend.position = "top")
```

The default amount of jitter is a little too much for our purposes. We can control it using height and width arguments to a position_jitter() function within the geom. Because we're making a one-dimensional summary here, we just need width. Figure 5.12 shows the data with a more appropriate amount of jitter.

Can you see why we did not use height? If not, try it and see what happens.

```
p ← ggplot(data = organdata,
           mapping = aes(x = reorder(country, donors, na.rm=TRUE),
                         y = donors, color = world))
p + geom_jitter(position = position_jitter(width=0.15)) +
   labs(x=NULL) + coord_flip() + theme(legend.position = "top")
```

When we want to summarize a categorical variable that just has one point per category, we should use this approach as well. The result will be a Cleveland dotplot, a simple and extremely effective method of presenting data that is usually better than either a bar chart or a table. For example, we can make a Cleveland dotplot of the average donation rate.

This also gives us another opportunity to do a little bit of data munging with a dplyr pipeline. We will use one to aggregate our larger country-year data frame to a smaller table of summary statistics by country. There is more than one way to use a pipeline for this task. We could choose the variables we want to summarize and then repeatedly use the mean() and sd() functions to calculate the means and standard deviations of the variables we want.

```
by_country ← organdata %>% group_by(consent_law, country) %>%
    summarize(donors_mean= mean(donors, na.rm = TRUE),
            donors_sd = sd(donors, na.rm = TRUE),
            gdp_mean = mean(gdp, na.rm = TRUE),
            health_mean = mean(health, na.rm = TRUE),
            roads_mean = mean(roads, na.rm = TRUE),
            cerebvas_mean = mean(cerebvas, na.rm = TRUE))
```

The pipeline consists of two steps. First we group the data by consent_law and country, and then we use summarize() to create six new variables, each one of which is the mean or standard deviation of each country's score on a corresponding variable in the original organdata data frame.

For an alternative view, change country to year in the grouping statement and see what happens.

As usual, the summarize() step will inherit information about the original data and the grouping and then do its calculations at the innermost grouping level. In this case it takes all the observations for each country and calculates the mean or standard deviation as requested. Here is what the resulting object looks like:

```
by_country
```

```
## # A tibble: 17 x 8
## # Groups:   consent_law [?]
##    consent_law country        donors_mean donors_sd gdp_mean health_mean roads_mean cerebvas_mean
##    <chr>       <chr>                <dbl>     <dbl>    <dbl>       <dbl>      <dbl>         <dbl>
##  1 Informed    Australia             10.6      1.14   22179.       1958.       105.          558.
##  2 Informed    Canada                14.0      0.751  23711.       2272.       109.          422.
##  3 Informed    Denmark               13.1      1.47   23722.       2054.       102.          641.
##  4 Informed    Germany               13.0      0.611  22163.       2349.       113.          707.
##  5 Informed    Ireland               19.8      2.48   20824.       1480.       118.          705.
##  6 Informed    Netherlands           13.7      1.55   23013.       1993.        76.1         585.
##  7 Informed    United Kingdom        13.5      0.775  21359.       1561.        67.9         708.
##  8 Informed    United States         20.0      1.33   29212.       3988.       155.          444.
##  9 Presumed    Austria               23.5      2.42   23876.       1875.       150.          769.
## 10 Presumed    Belgium               21.9      1.94   22500.       1958.       155.          594.
```

## 11 Presumed	Finland	18.4	1.53	21019.	1615.	93.6	771.
## 12 Presumed	France	16.8	1.60	22603.	2160.	156.	433.
## 13 Presumed	Italy	11.1	4.28	21554.	1757.	122.	712.
## 14 Presumed	Norway	15.4	1.11	26448.	2217.	70.0	662.
## 15 Presumed	Spain	28.1	4.96	16933.	1289.	161.	655.
## 16 Presumed	Sweden	13.1	1.75	22415.	1951.	72.3	595.
## 17 Presumed	Switzerland	14.2	1.71	27233.	2776.	96.4	424.

As before, the variables specified in group_by() are retained in the new data frame, the variables created with summarize() are added, and all the other variables in the original data are dropped. The countries are also summarized alphabetically within consent_law, which was the outermost grouping variable in the group_by() statement at the start of the pipeline.

Using our pipeline this way is reasonable, but the code is worth looking at again. For one thing, we have to repeatedly type out the names of the mean() and sd() functions and give each of them the name of the variable we want summarized *and* the na.rm = TRUE argument each time to make sure the functions don't complain about missing values. We also repeatedly name our new summary variables in the same way, by adding _mean or _sd to the end of the original variable name. If we wanted to calculate the mean and standard deviation for all the numerical variables in organdata, our code would get even longer. Plus, in this version we lose the other, time-invariant categorical variables that we haven't grouped by, such as world. When we see repeated actions like this in our code, we can ask whether there's a better way to proceed.

There is. What we would like to do is apply the mean() and sd() functions to every numerical variable in organdata, but *only* the numerical ones. Then we want to name the results in a consistent way and return a summary table including all the categorical variables like world. We can create a better version of the by_country object using a little bit of R's functional programming abilities. Here is the code:

```
by_country ← organdata %>% group_by(consent_law, country) %>%
    summarize_if(is.numeric, funs(mean, sd), na.rm = TRUE) %>%
    ungroup()
```

The pipeline starts off just as before, taking organdata and then grouping it by consent_law and country. In the next step, though, instead of manually taking the mean and standard

We do not have to use parentheses when naming the functions inside summarize_if().

deviation of a subset of variables, we use the summarize_if() function instead. As its name suggests, it examines each column in our data and applies a test to it. It only summarizes if the test is passed, that is, if it returns a value of TRUE. Here the test is the function is.numeric(), which looks to see if a vector is a numeric value. If it is, then summarize_if() will apply the summary function or functions we want to organdata. Because we are taking both the mean and the standard deviation, we use funs() to list the functions we want. And we finish with the na.rm = TRUE argument, which will be passed on to each use of both mean() and sd(). In the last step in the pipeline we ungroup() the data, so that the result is a plain tibble.

Sometimes graphing functions can get confused by grouped tibbles where we don't explicitly use the groups in the plot.

Here is what the pipeline returns:

by_country

```
## # A tibble: 17 x 28
##    consent_law country      donors_mean pop_mean pop_dens_mean gdp_mean gdp_lag_mean health_mean
##    <chr>       <chr>              <dbl>    <dbl>         <dbl>    <dbl>        <dbl>       <dbl>
##  1 Informed    Australia           10.6   18318.         0.237   22179.       21779.       1958.
##  2 Informed    Canada              14.0   29608.         0.297   23711.       23353.       2272.
##  3 Informed    Denmark             13.1    5257.        12.2     23722.       23275.       2054.
##  4 Informed    Germany             13.0   80255.        22.5     22163.       21938.       2349.
##  5 Informed    Ireland             19.8    3674.         5.23    20824.       20154.       1480.
##  6 Informed    Netherlands         13.7   15548.        37.4     23013.       22554.       1993.
##  7 Informed    United Kingdom      13.5   58187.        24.0     21359.       20962.       1561.
##  8 Informed    United States       20.0  269330.         2.80    29212.       28699.       3988.
##  9 Presumed    Austria             23.5    7927.         9.45    23876.       23415.       1875.
## 10 Presumed    Belgium             21.9   10153.        30.7     22500.       22096.       1958.
## 11 Presumed    Finland             18.4    5112.         1.51    21019.       20763.       1615.
## 12 Presumed    France              16.8   58056.        10.5     22603.       22211.       2160.
## 13 Presumed    Italy               11.1   57360.        19.0     21554.       21195.       1757.
## 14 Presumed    Norway              15.4    4386.         1.35    26448.       25769.       2217.
## 15 Presumed    Spain               28.1   39666.         7.84    16933.       16584.       1289.
## 16 Presumed    Sweden              13.1    8789.         1.95    22415.       22094.       1951.
## 17 Presumed    Switzerland         14.2    7037.        17.0     27233.       26931.       2776.
## # ... with 20 more variables: health_lag_mean <dbl>, pubhealth_mean <dbl>, roads_mean <dbl>,
## #   cerebvas_mean <dbl>, assault_mean <dbl>, external_mean <dbl>, txp_pop_mean <dbl>,
## #   donors_sd <dbl>, pop_sd <dbl>, pop_dens_sd <dbl>, gdp_sd <dbl>, gdp_lag_sd <dbl>,
## #   health_sd <dbl>, health_lag_sd <dbl>, pubhealth_sd <dbl>, roads_sd <dbl>, cerebvas_sd <dbl>,
## #   assault_sd <dbl>, external_sd <dbl>, txp_pop_sd <dbl>
```

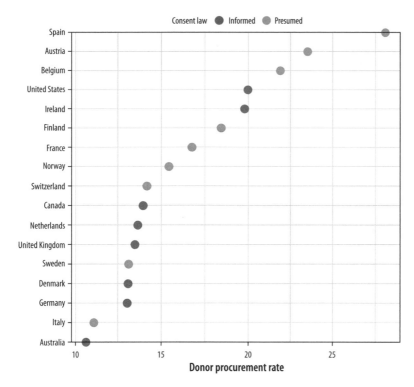

Figure 5.13: A Cleveland dotplot, with colored points.

All the numeric variables have been summarized. They are named using the original variable, with the function's name appended: donors_mean and donors_sd, and so on. This is a compact way to rapidly transform our data in various ways. There is a family of summarize_ functions for various tasks, and a complementary group of mutate_ functions for when we want to add columns to the data rather than aggregate it.

With our data summarized by country, for figure 5.13 we can draw a dotplot with geom_point(). Let's also color the results by the consent law for each country.

```
p ← ggplot(data = by_country,
           mapping = aes(x = donors_mean, y = reorder(country, donors_mean),
                         color = consent_law))
p + geom_point(size=3) +
    labs(x = "Donor Procurement Rate",
         y = "", color = "Consent Law") +
    theme(legend.position="top")
```

Alternatively, if we liked, we could use a facet instead of coloring the points. Using `facet_wrap()`, we can split the `consent_law` variable into two panels and then rank the countries by donation rate within each panel. Because we have a categorical variable on our y-axis, there are two wrinkles worth noting. First, if we leave `facet_wrap()` to its defaults, the panels will be plotted side by side. This will make it difficult to compare the two groups on the same scale. Instead the plot will be read left to right, which is not useful. To avoid this, we will have the panels appear one on top of the other by saying we want to have only one column. This is the `ncol=1` argument. Second, and again because we have a categorical variable on the y-axis, the default facet plot will have the names of every country appear on the y-axis of *both* panels. (Were the y-axis a continuous variable, this would be what we would want.) In that case, only half the rows in each panel of our plot will have points in them.

To avoid this we allow the y-axes scale to be free. This is the `scales = "free_y"` argument. Again, for faceted plots where both variables are continuous, we generally do not want the scales to be free, because it allows the x- or y-axis for each panel to vary with the range of the data inside that panel only, instead of the range across the whole dataset. Ordinarily, the point of small-multiple facets is to be able to compare across the panels. This means free scales are usually not a good idea, because each panel gets its own x- or y-axis range, which breaks comparability. But where one axis is categorical, as in figure 5.14, we can free the categorical axis and leave the continuous one fixed. The result is that each panel shares the same x-axis, and it is easy to compare them.

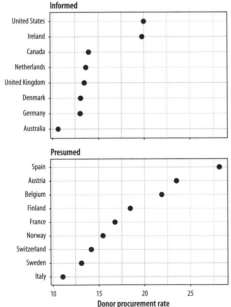

Figure 5.14: A faceted dotplot with free scales on the y-axis.

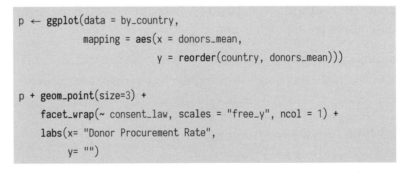

```
p ← ggplot(data = by_country,
           mapping = aes(x = donors_mean,
                         y = reorder(country, donors_mean)))


p + geom_point(size=3) +
    facet_wrap(~ consent_law, scales = "free_y", ncol = 1) +
    labs(x= "Donor Procurement Rate",
         y= "")
```

Cleveland dotplots are generally preferred to bar or column charts. When making them, put the categories on the y-axis and order them in the way that is most relevant to the numerical summary you are providing. This sort of plot is also an excellent way to summarize model results or any data with with error ranges. We use geom_point() to draw our dotplots. There is a geom called geom_dotplot(), but it is designed to produce a different sort of figure. It is a kind of histogram, with individual observations represented by dots that are then stacked on top of one another to show how many of them there are.

The Cleveland-style dotplot can be extended to cases where we want to include some information about variance or error in the plot. Using geom_pointrange(), we can tell ggplot to show us a point estimate and a range around it. Here we will use the standard deviation of the donation rate that we calculated above. But this is also the natural way to present, for example, estimates of model coefficients with confidence intervals. With geom_pointrange() we map our x and y variables as usual, but the function needs a little more information than geom_point. It needs to know the range of the line to draw on either side of the point, defined by the arguments ymax and ymin. This is given by the y value (donors_mean) plus or minus its standard deviation (donors_sd). If a function argument expects a number, it is OK to give it a mathematical expression that resolves to the number you want. R will calculate the result for you.

```
p ← ggplot(data = by_country, mapping = aes(x = reorder(country,
           donors_mean), y = donors_mean))


p + geom_pointrange(mapping = aes(ymin = donors_mean - donors_sd,
       ymax = donors_mean + donors_sd)) +
    labs(x= "", y= "Donor Procurement Rate") + coord_flip()
```

Because geom_pointrange() expects y, ymin, and ymax as arguments, in figure 5.15 we map donors_mean to y and the ccode variable to x, then flip the axes at the end with coord_flip().

5.3 Plot Text Directly

It can sometimes be useful to plot the labels along with the points in a scatterplot, or just plot informative labels directly. We can do this with geom_text().

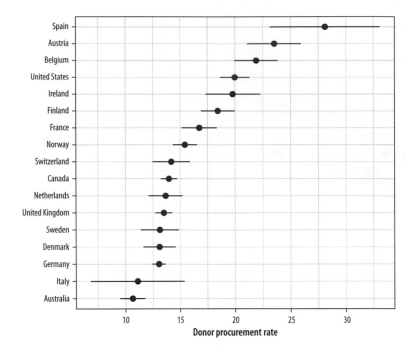

Figure 5.15: A dot-and-whisker plot, with the range defined by the standard deviation of the measured variable.

```
p ← ggplot(data = by_country, mapping = aes(x = roads_mean,
    y = donors_mean))
p + geom_point() + geom_text(mapping = aes(label = country))
```

The text in figure 5.16 is plotted right on top of the points because both are positioned using the same x and y mapping. One way of dealing with this, often the most effective if we are not too worried about excessive precision in the graph, is to remove the points by dropping geom_point() from the plot. A second option is to adjust the position of the text. We can left- or right-justify the labels using the hjust argument to geom_text(). Setting hjust = 0 will left-justify the label, and hjust=1 will right-justify it.

```
p ← ggplot(data = by_country,
           mapping = aes(x = roads_mean, y = donors_mean))

p + geom_point() + geom_text(mapping = aes(label = country), hjust = 0)
```

You might be tempted to try different values to hjust to fine-tune the labels, in figure 5.17, but this is not a robust approach. It will often fail because the space is added in proportion to the length of the label. The result is that longer labels move further

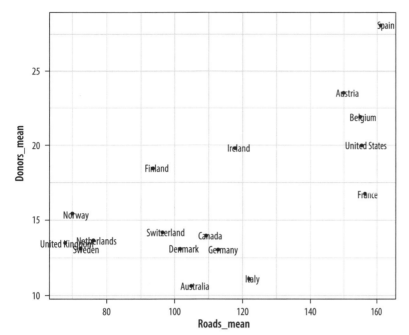

Figure 5.16: Plotting labels and text.

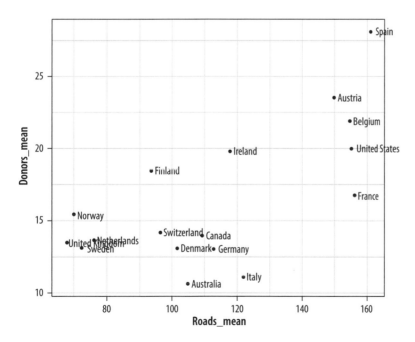

Figure 5.17: Plot points and text labels, with a horizontal position adjustment.

away from their points than you want. There are ways around this, but they introduce other problems.

Instead of wrestling any further with geom_text(), we will use ggrepel. This useful package adds new geoms to ggplot. Just as ggplot extends the plotting capabilities of R, there are many small packages that extend the capabilities of ggplot, often by providing some new type of geom. The ggrepel package provides geom_text_repel() and geom_label_repel(), two geoms that can pick out labels much more flexibly than the default geom_text(). First, make sure the library is installed, then load it in the usual way:

```
library(ggrepel)
```

We will use geom_text_repel() instead of geom_text(). To demonstrate some of what geom_text_repel() can do, we will switch datasets and work with some historical U.S. presidential election data provided in the socviz library.

```
elections_historic %>% select(2:7)
```

```
## # A tibble: 49 x 6
##    year winner                   win_party ec_pct popular_pct popular_margin
##    <int> <chr>                   <chr>     <dbl>  <dbl>       <dbl>
## 1  1824 John Quincy Adams        D.-R.     0.322  0.309       -0.104
## 2  1828 Andrew Jackson           Dem.      0.682  0.559       0.122
## 3  1832 Andrew Jackson           Dem.      0.766  0.547       0.178
## 4  1836 Martin Van Buren         Dem.      0.578  0.508       0.142
## 5  1840 William Henry Harrison   Whig      0.796  0.529       0.0605
## 6  1844 James Polk               Dem.      0.618  0.495       0.0145
## 7  1848 Zachary Taylor           Whig      0.562  0.473       0.0479
## 8  1852 Franklin Pierce          Dem.      0.858  0.508       0.0695
## 9  1856 James Buchanan           Dem.      0.588  0.453       0.122
## 10 1860 Abraham Lincoln          Rep.      0.594  0.396       0.101
## # ... with 39 more rows
```

```
p_title ← "Presidential Elections: Popular & Electoral College Margins"
p_subtitle ← "1824-2016"
p_caption ← "Data for 2016 are provisional."
```

```
x_label ← "Winner's share of Popular Vote"
y_label ← "Winner's share of Electoral College Votes"


p ← ggplot(elections_historic, aes(x = popular_pct, y = ec_pct,
                                   label = winner_label))


p + geom_hline(yintercept = 0.5, size = 1.4, color = "gray80") +
    geom_vline(xintercept = 0.5, size = 1.4, color = "gray80") +
    geom_point() +
    geom_text_repel() +
    scale_x_continuous(labels = scales::percent) +
    scale_y_continuous(labels = scales::percent) +
    labs(x = x_label, y = y_label, title = p_title, subtitle = p_subtitle,
         caption = p_caption)
```

Figure 5.18 takes each U.S. presidential election since 1824 (the first year that the size of the popular vote was recorded) and plots the winner's share of the popular vote against the winner's share of the electoral college vote. The shares are stored in the data as proportions (from 0 to 1) rather than percentages, so we need to adjust the labels of the scales using scale_x_continuous() and scale_y_continuous(). Seeing as we are interested in particular presidencies, we also want to label the points. But because many of the data points are plotted quite close together, we need to make sure the labels do not overlap, or obscure other points. The geom_text_repel() function handles the problem very well. This plot has relatively long titles. We could put them directly in the code, but to keep things tidier we assign the text to some named objects instead. Then we use those in the plot formula.

In this plot, what is of interest about any particular point is the quadrant of the x-y plane each point is in, and how far away it is from the 50 percent threshold on both the x-axis (with the popular vote share) and the y-axis (with the Electoral College vote share). To underscore this point we draw two reference lines at the 50 percent line in each direction. They are drawn at the beginning of the plotting process so that the points and labels can be layered on top of them. We use two new geoms, geom_hline() and geom_vline(), to make the lines. They take yintercept and xintercept arguments, respectively, and the lines can also be sized and colored as you please. There is also a geom_abline() geom

Normally it is not a good idea to label every point on a plot in the way we do here. A better approach might be to select a few points of particular interest.

Presidential elections: Popular & electoral college margins

1824–2016

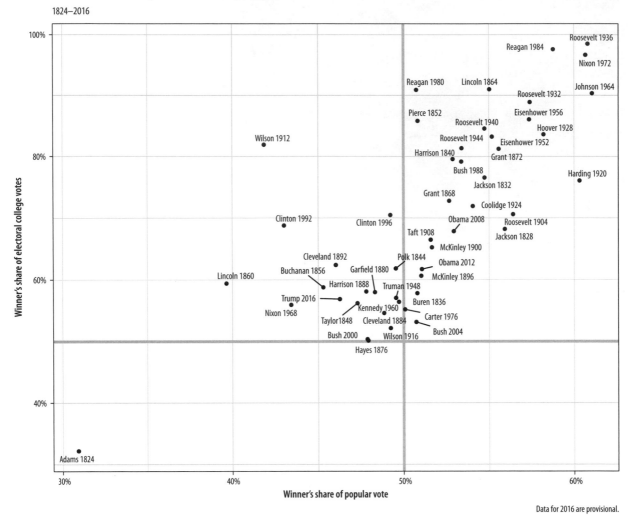

Figure 5.18: Text labels with ggrepel.

that draws straight lines based on a supplied slope and intercept. This is useful for plotting, for example, 45 degree reference lines in scatterplots.

The ggrepel package has several other useful geoms and options to aid with effectively plotting labels along with points. The performance of its labeling algorithm is consistently very good. For most purposes it will be a better first choice than geom_text().

5.4 Label Outliers

Sometimes we want to pick out some points of interest in the data without labeling every single item. We can still use geom_text() or geom_text_repel(). We just need to select the points we want to label. We do this by telling geom_text_repel() to use a different dataset from the one geom_point() is using. The subset() function does the work.

```
p ← ggplot(data = by_country,
           mapping = aes(x = gdp_mean, y = health_mean))

p + geom_point() +
    geom_text_repel(data = subset(by_country, gdp_mean > 25000),
                    mapping = aes(label = country))

p ← ggplot(data = by_country,
           mapping = aes(x = gdp_mean, y = health_mean))

p + geom_point() +
    geom_text_repel(data = subset(by_country,
                                  gdp_mean > 25000 | health_mean < 1500 |
                                  country %in% "Belgium"),
                    mapping = aes(label = country))
```

In the top part of figure 5.19, we specify a new data argument to the text geom and use subset() to create a small dataset on the fly. The subset() function takes the by_country object and selects only the cases where gdp_mean is over 25,000, with the result that only those points are labeled in the plot. The criteria we use can be whatever we like, as long as we can write a logical expression that defines it. For example, in the lower part of the figure we pick out cases where gdp_mean is greater than 25,000, *or* health_mean is less than 1,500, *or* the country is Belgium. In all these plots, because we are using geom_text_repel(), we no longer have to worry about our earlier problem where the country labels were clipped at the edge of the plot.

Alternatively, we can pick out specific points by creating a dummy variable in the data set just for this purpose. For figure 5.20 we add a column to organdata called ind. An observation gets coded as TRUE if ccode is "Ita" or "Spa," *and* if the year is greater

Figure 5.19: *Top*: Labeling text according to a single criterion. *Bottom*: Labeling according to several criteria.

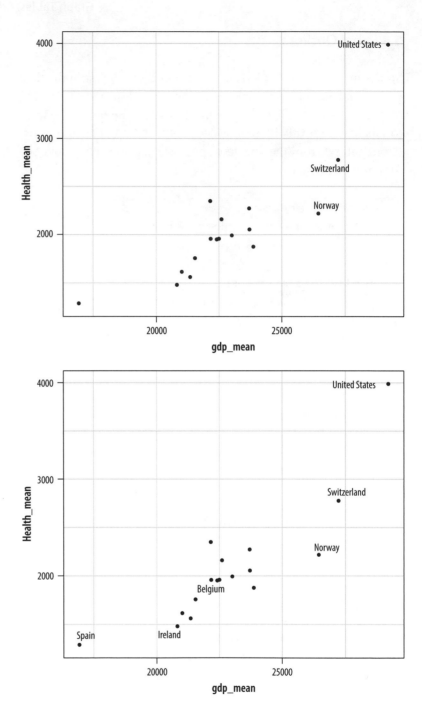

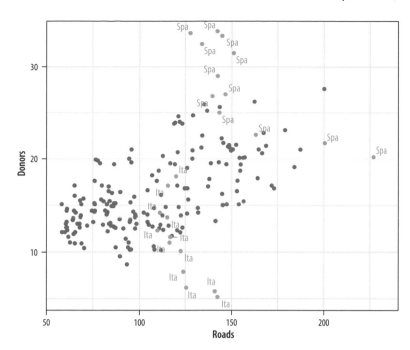

Figure 5.20: Labeling using a dummy variable.

than 1998. We use this new `ind` variable in two ways in the plotting code. First, we map it to the `color` aesthetic in the usual way. Second, we use it to subset the data that the text geom will label. Then we suppress the legend that would otherwise appear for the `label` and `color` aesthetics by using the `guides()` function.

```r
organdata$ind ← organdata$ccode %in% c("Ita", "Spa") &
                    organdata$year > 1998


p ← ggplot(data = organdata,
           mapping = aes(x = roads,
                         y = donors, color = ind))
p + geom_point() +
    geom_text_repel(data = subset(organdata, ind),
                    mapping = aes(label = ccode)) +
    guides(label = FALSE, color = FALSE)
```

5.5 Write and Draw in the Plot Area

Sometimes we want to annotate the figure directly. Maybe we need to point out something important that is not mapped to a variable. We use `annotate()` for this purpose. It isn't quite a geom, as it doesn't accept any variable mappings from our data. Instead, it can *use* geoms, temporarily taking advantage of their features in order to place something on the plot. The most obvious use-case is putting arbitrary text on the plot (fig. 5.21).

We will tell `annotate()` to use a text geom. It hands the plotting duties to `geom_text()`, which means that we can use all of that geom's arguments in the `annotate()` call. This includes the x, y, and `label` arguments, as one would expect, but also things like `size`, `color`, and the `hjust` and `vjust` settings that allow text to be justified. This is particularly useful when our label has several lines in it. We include extra lines by using the special "newline" code, \n, which we use instead of a space to force a line-break as needed.

```
p ← ggplot(data = organdata, mapping = aes(x = roads, y = donors))
p + geom_point() + annotate(geom = "text", x = 91, y = 33,
                            label = "A surprisingly high \n recovery rate.",
                            hjust = 0)
```

The `annotate()` function can work with other geoms, too. Use it to draw rectangles, line segments, and arrows. Just remember to pass along the right arguments to the geom you use. We can add a rectangle to this plot, (fig 5.22), for instance, with a second call to the function.

```
p ← ggplot(data = organdata,
           mapping = aes(x = roads, y = donors))
p + geom_point() +
    annotate(geom = "rect", xmin = 125, xmax = 155,
             ymin = 30, ymax = 35, fill = "red", alpha = 0.2) +
    annotate(geom = "text", x = 157, y = 33,
             label = "A surprisingly high \n recovery rate.", hjust = 0)
```

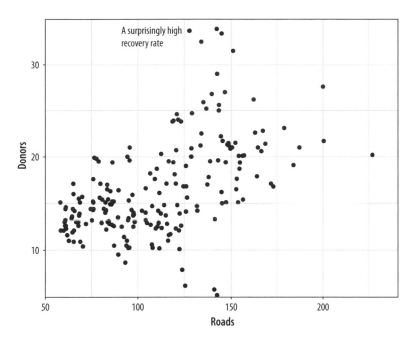

A surprisingly high
recovery rate

Figure 5.21: Arbitrary text with annotate().

5.6 Understanding Scales, Guides, and Themes

This chapter has gradually extended our ggplot vocabulary in two ways. First, we introduced some new geom_ functions that allowed us to draw new kinds of plots. Second, we made use of new functions controlling some aspects of the appearance of our graph. We used scale_x_log10(), scale_x_continuous(), and other scale_ functions to adjust axis labels. We used the guides() function to remove the legends for a color mapping and a label mapping. And we also used the theme() function to move the position of a legend from the side to the top of a figure.

Learning about new geoms extended what we have seen already. Each geom makes a different type of plot. Different plots require different mappings in order to work, and so each geom_ function takes mappings tailored to the kind of graph it draws. You can't use geom_point() to make a scatterplot without supplying an x and a y mapping, for example. Using geom_histogram() only requires you to supply an x mapping. Similarly, geom_pointrange() requires ymin and ymax mappings in order to know where to draw the line ranges it makes. A geom_ function may take optional arguments, too. When using geom_boxplot() you can specify what the outliers look like using arguments like outlier.shape and outlier.color.

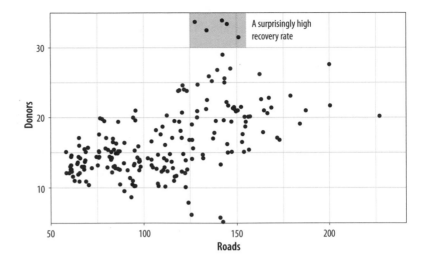

Figure 5.22: Using two different geoms with annotate().

The second kind of extension introduced some new functions, and with them some new concepts. What are the differences between the scale_ functions, the guides() function, and the theme() function? When do you know to use one rather than the other? Why are there so many scale_ functions listed in the online help, anyway? How can you tell which one you need?

Here is a rough starting point:

- Every aesthetic mapping has a scale. If you want to adjust how that scale is marked or graduated, then you use a scale_ function.

- Many scales come with a legend or key to help the reader interpret the graph. These are called *guides*. You can make adjustments to them with the guides() function. Perhaps the most common use case is to make the legend disappear, as sometimes it is superfluous. Another is to adjust the arrangement of the key in legends and color bars.

- Graphs have other features not strictly connected to the logical structure of the data being displayed. These include things like their background color, the typeface used for labels, or the placement of the legend on the graph. To adjust these, use the theme() function.

Consistent with ggplot's overall approach, adjusting some visible feature of the graph means first thinking about the relationship the feature has with the underlying data. Roughly speaking, if the change you want to make will affect the substantive interpretation of any particular geom, then most likely you will either be mapping an aesthetic to a variable using that geom's aes() function or be specifying a change via some scale_ function. If the change you want to make does not affect the interpretation of a given geom_, then most likely you will either be setting a variable inside the geom_ function, or making a cosmetic change via the theme() function.

Scales and guides are closely connected, which can make things confusing. The guide provides information about the scale, such as in a legend or color bar. Thus it is possible to make adjustments to guides from inside the various scale_ functions. More often it is easier to use the guides() function directly.

```
p ← ggplot(data = organdata,
           mapping = aes(x = roads,
                         y = donors,
                         color = world))
p + geom_point()
```

Figure 5.23 shows a plot with three aesthetic mappings. The variable roads is mapped to x; donors is mapped to y; and world is mapped to color. The x and y scales are both *continuous*, running smoothly from just under the lowest value of the variable to just over the highest value. Various labeled tick-marks orient the reader to the values on each axis. The color mapping also has a scale. The world measure is an unordered categorical variable, so its scale is *discrete*. It takes one of four values, each represented by a different color.

Along with color, mappings like fill, shape, and size will have scales that we might want to customize or adjust. We could have mapped world to shape instead of color. In that case our four-category variable would have a scale consisting of four different shapes. Scales for these mappings may have labels, axis tick-marks at particular positions, or specific colors or shapes. If we want to adjust them, we use one of the scale_ functions.

Many different kinds of variable can be mapped. More often than not, x and y are continuous measures. But they might also easily be discrete, as when we mapped country names to the y-axis in our boxplots and dotplots. An x or y mapping can also be defined as a transformation onto a log scale, or as a special sort of number value like a date. Similarly, a color or a fill mapping

Figure 5.23: Every mapped variable has a scale.

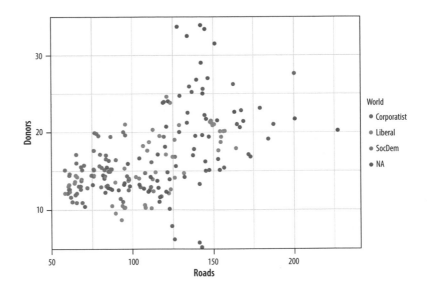

scale_<MAPPING>_<KIND>()

Figure 5.24: A schema for naming the scale functions.

can be discrete and *unordered*, as with our world variable, or discrete and *ordered*, as with letter grades in an exam. A color or fill mapping can also be a continuous quantity, represented as a gradient running smoothly from a low to a high value. Finally, both continuous gradients and ordered discrete values might have some defined neutral midpoint with extremes diverging in both directions.

Because we have several potential mappings, and each mapping might be to one of several different scales, we end up with a lot of individual scale_ functions. Each deals with one combination of mapping and scale. They are named according to a consistent logic, shown in figure 5.24. First comes the scale_ name, then the *mapping* it applies to, and finally the *kind* of value the scale will display. Thus the scale_x_continuous() function controls x scales for continuous variables; scale_y_discrete() adjusts y scales for discrete variables; and scale_x_log10() transforms an x mapping to a log scale. Most of the time, ggplot will guess correctly what sort of scale is needed for your mapping. Then it will work out some default features of the scale (such as its labels and where the tick-marks go). In many cases you will not need to make any scale adjustments. If x is mapped to a continuous variable, then adding + scale_x_continuous() to your plot statement with no further arguments will have no effect. It is already there implicitly. Adding + scale_x_log10(), on the other hand, will transform

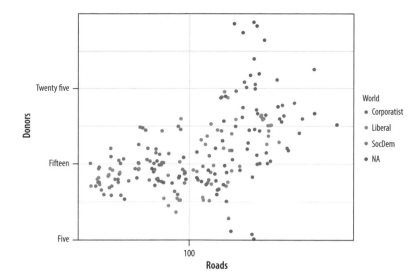

your scale, as now you have replaced the default treatment of a continuous x variable.

If you want to adjust the labels or tick-marks on a scale, you will need to know which mapping it is for and what sort of scale it is. Then you supply the arguments to the appropriate scale function. For example, we can change the x-axis of the previous plot to a log scale and then also change the position and labels of the tick-marks on the y-axis (fig. 5.25).

```
p ← ggplot(data = organdata, mapping = aes(x = roads, y = donors,
    color = world))
p + geom_point() + scale_x_log10() + scale_y_continuous(breaks = c(5,
    15, 25), labels = c("Five", "Fifteen", "Twenty Five"))
```

The same applies to mappings like color and fill (see fig. 5.26). Here the available scale_ functions include ones that deal with continuous, diverging, and discrete variables, as well as others that we will encounter later when we discuss the use of color and color palettes in more detail. When working with a scale that produces a legend, we can also use its scale_ function to specify the labels in the key. To change the *title* of the legend, however, we use the labs() function, which lets us label all the mappings.

```
p ← ggplot(data = organdata, mapping = aes(x = roads, y = donors,
    color = world))
```

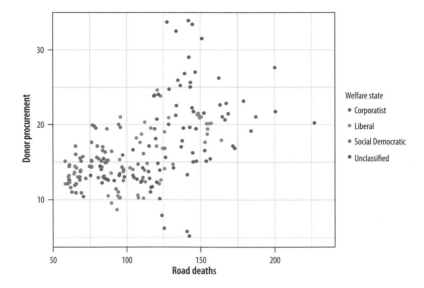

```
p + geom_point() + scale_color_discrete(labels = c("Corporatist",
    "Liberal", "Social Democratic", "Unclassified")) + labs(x = "Road Deaths",
    y = "Donor Procurement", color = "Welfare State")
```

If we want to move the legend somewhere else on the plot, we are making a purely cosmetic decision and that is the job of the theme() function. As we have already seen, adding + theme(legend.position = "top") will move the legend as instructed. Finally, to make the legend disappear altogether (fig. 5.27), we tell ggplot that we do not want a guide for that scale. This is generally not good practice, but there can be reasons to do it. We already saw an example in figure 4.9.

```
p ← ggplot(data = organdata, mapping = aes(x = roads, y = donors,
    color = world))
p + geom_point() + labs(x = "Road Deaths", y = "Donor Procurement") +
    guides(color = FALSE)
```

We will look more closely at scale_ and theme() functions in chapter 8, when we discuss how to polish plots that we are ready to display or publish. Until then, we will use scale_ functions fairly regularly to make small adjustments to the labels and axes of our graphs. And we will occasionally use the theme() function

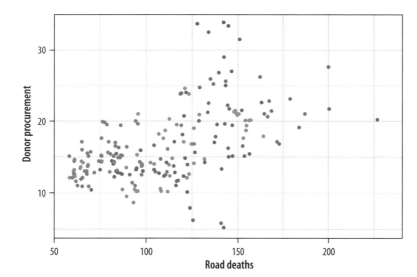

Figure 5.27: Removing the guide to a scale.

to make some cosmetic adjustments. So you do not need to worry about additional details of how they work until later on. But at this point it *is* worth knowing what scale_ functions are for, and the logic behind their naming scheme. Understanding the scale_<mapping>_<kind>() rule makes it easier to see what is going on when one of these functions is called to make an adjustment to a plot.

5.7 Where to Go Next

We covered several new functions and data aggregation techniques in this chapter. You should practice working with them.

- The subset() function is very useful when used in conjunction with a series of layered geoms. Go back to your code for the presidential elections plot (fig. 5.18) and redo it so that it shows all the data points but only labels elections since 1992. You might need to look again at the elections_historic data to see what variables are available to you. You can also experiment with subsetting by political party, or changing the colors of the points to reflect the winning party.
- Use geom_point() and reorder() to make a Cleveland dotplot of all presidential elections, ordered by share of the popular vote.

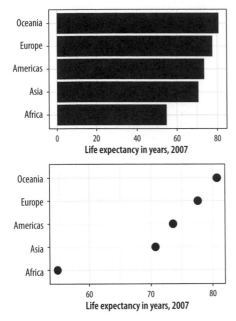

Figure 5.28: Two figures from chapter 1.

- Try using annotate() to add a rectangle that lightly colors the entire upper left quadrant of figure 5.18.
- The main action verbs in the dplyr library are group_by(), filter(), select(), summarize(), and mutate(). Practice with them by revisiting the gapminder data to see if you can reproduce a pair of graphs from chapter 1, shown here again in figure 5.28. You will need to filter some rows, group the data by continent, and calculate the mean life expectancy by continent before beginning the plotting process.
- Get comfortable with grouping, mutating, and summarizing data in pipelines. This will become a routine task as you work with your data. There are many ways that tables can be aggregated and transformed. Remember, group_by() groups your data from left to right, with the rightmost or innermost group being the level calculations will be done at; mutate() adds a column at the current level of grouping; and summarize() aggregates to the next level up. Try creating some grouped objects from the GSS data, calculating frequencies as you learned in this chapter, and then check to see if the totals are what you expect. For example, start by grouping degree by race, like this:

```
gss_sm %>% group_by(race, degree) %>% summarize(N = n()) %>%
    mutate(pct = round(N/sum(N) * 100, 0))
```

```
## # A tibble: 18 x 4
## # Groups:   race [3]
##    race  degree              N   pct
##    <fct> <fct>           <int> <dbl>
##  1 White Lt High School    197    9.
##  2 White High School      1057   50.
##  3 White Junior College    166    8.
##  4 White Bachelor          426   20.
##  5 White Graduate          250   12.
##  6 White <NA>                4    0.
##  7 Black Lt High School     60   12.
##  8 Black High School       292   60.
##  9 Black Junior College     33    7.
## 10 Black Bachelor           71   14.
## 11 Black Graduate           31    6.
```

```
## 12 Black <NA>             3    1.
## 13 Other Lt High School   71   26.
## 14 Other High School     112   40.
## 15 Other Junior College   17    6.
## 16 Other Bachelor         39   14.
## 17 Other Graduate         37   13.
## 18 Other <NA>              1    0.
```

- This code is similar to what you saw earlier but more compact. (We calculate the `pct` values directly.) Check that the results are as you expect by grouping by `race` and summing the percentages. Try doing the same exercise grouping by `sex` or `region`.

- Try summary calculations with functions other than `sum`. Can you calculate the mean and median number of children by `degree`? (Hint: the `childs` variable in `gss_sm` has children as a numeric value.)

- `dplyr` has a large number of helper functions that let you summarize data in many different ways. The vignette on *window functions* included with the `dplyr` documentation is a good place to begin learning about these. You should also look at chapter 3 of Wickham & Grolemund (2016) for more information on transforming data with `dplyr`.

- Experiment with the `gapminder` data to practice some of the new geoms we have learned. Try examining population or life expectancy over time using a series of boxplots. (Hint: you may need to use the `group` aesthetic in the `aes()` call.) Can you facet this boxplot by continent? Is anything different if you create a tibble from `gapminder` that explicitly groups the data by `year` and `continent` first, and then create your plots with that?

- Read the help page for `geom_boxplot()` and take a look at the `notch` and `varwidth` options. Try them out to see how they change the look of the plot.

- As an alternative to `geom_boxplot()`, try `geom_violin()` for a similar plot but with a mirrored density distribution instead of a box and whiskers.

- `geom_pointrange()` is one of a family of related geoms that produce different kinds of error bars and ranges, depending on your specific needs. They include `geom_linerange()`, `geom_crossbar()`, and `geom_errorbar()`. Try them out using `gapminder` or `organdata` to see how they differ.

6 Work with Models

Data visualization is about more than generating figures that display the raw numbers from a table of data. Right from the beginning, it involves summarizing or transforming parts of the data and then plotting the results. Statistical models are a central part of that process. In this chapter, we will begin by looking briefly at how ggplot can use various modeling techniques directly within geoms. Then we will see how to use the broom and margins libraries to tidily extract and plot estimates from models that we fit ourselves.

```r
p ← ggplot(data = gapminder,
           mapping = aes(x = log(gdpPercap), y = lifeExp))


p + geom_point(alpha=0.1) +
    geom_smooth(color = "tomato", fill="tomato", method = MASS::rlm) +
    geom_smooth(color = "steelblue", fill="steelblue", method = "lm")


p + geom_point(alpha=0.1) +
    geom_smooth(color = "tomato", method = "lm", size = 1.2,
                formula = y ~ splines::bs(x, 3), se = FALSE)


p + geom_point(alpha=0.1) +
    geom_quantile(color = "tomato", size = 1.2, method = "rqss",
                  lambda = 1, quantiles = c(0.20, 0.5, 0.85))
```

Histograms, density plots, boxplots, and other geoms compute either single numbers or new variables before plotting them. As we saw in section 4.4, these calculations are done by stat_ functions, each of which works hand in hand with its default geom_ function, and vice versa. Moreover, from the smoothing lines we drew in almost the very first plots we made, we have seen that stat_ functions can do a fair amount of calculation and even model estimation directly. The geom_smooth() function can take a range of method arguments to fit LOESS, OLS, and robust regression lines, among others.

Both the geom_smooth() and geom_quantile() functions can also be instructed to use different formulas to produce their fits. In the top panel of figure 6.1, we access the MASS library's rlm function to fit a robust regression line. In the second panel, the bs function is invoked directly from the splines library in the same way, to fit a polynominal curve to the data. This is the same approach to directly accessing functions without loading a whole library that we have already used several times when using functions from the scales package. The geom_quantile() function, meanwhile, is like a specialized version of geom_smooth() that can fit quantile regression lines using a variety of methods. The quantiles argument takes a vector specifying the quantiles at which to fit the lines.

6.1 Show Several Fits at Once, with a Legend

As we just saw in the first panel of figure 6.1, where we plotted both an OLS and a robust regression line, we can look at several fits at once on the same plot by layering on new smoothers with geom_smooth(). As long as we set the color and fill aesthetics to different values for each fit, we can easily distinguish them visually. However, ggplot will not draw a legend that guides us about which fit is which. This is because the smoothers are not logically connected to one another. They exist as separate layers. What if we are comparing several different fits and want a legend describing them?

As it turns out, geom_smooth() can do this via the slightly unusual route of mapping the color and fill aesthetics to a string describing the model we are fitting and then using scale_color_manual() and scale_fill_manual() to create the legend (fig.6.2). First we use brewer.pal() from the RColor-Brewer library to extract three qualitatively different colors from a larger palette. The colors are represented as hex values. As before use the :: convention to use the function without loading the whole package:

```
model_colors ← RColorBrewer::brewer.pal(3, "Set1")
model_colors
```

```
## [1] "#E41A1C" "#377EB8" "#4DAF4A"
```

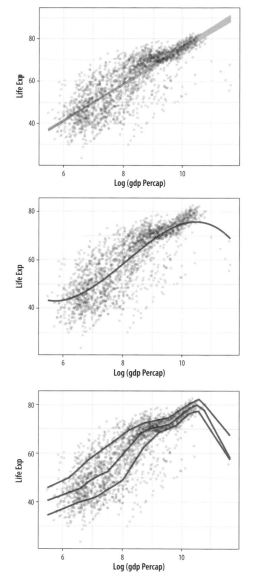

Figure 6.1: From top to bottom: an OLS vs robust regression comparison; a polynomial fit; and quantile regression.

Then we create a plot with three different smoothers, mapping the color and fill *within the aes() function* as the name of the smoother:

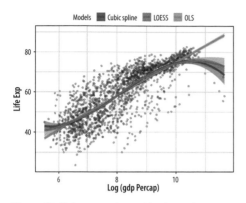

Figure 6.2: Fitting smoothers with a legend.

```
p0 ← ggplot(data = gapminder,
             mapping = aes(x = log(gdpPercap), y = lifeExp))


p1 ← p0 + geom_point(alpha = 0.2) +
    geom_smooth(method = "lm", aes(color = "OLS", fill = "OLS")) +
    geom_smooth(method = "lm", formula = y ~ splines::bs(x, df = 3),
                aes(color = "Cubic Spline", fill = "Cubic Spline")) +
    geom_smooth(method = "loess",
                aes(color = "LOESS", fill = "LOESS"))


p1 + scale_color_manual(name = "Models", values = model_colors) +
    scale_fill_manual(name = "Models", values = model_colors) +
    theme(legend.position = "top")
```

In a way we have cheated a little here to make the plot work. Until now, we have always mapped aesthetics to the names of variables, not to strings like "OLS" or "Cubic Splines." In chapter 3, when we discussed mapping versus setting aesthetics, we saw what happened when we tried to change the color of the points in a scatterplot by setting them to "purple" inside the aes() function. The result was that the points turned red instead, as ggplot in effect created a new variable and labeled it with the word "purple." We learned there that the aes() function was for mapping variables to aesthetics.

Here we take advantage of that behavior, creating a new single-value variable for the name of each of our models. Ggplot will properly construct the relevant guide if we call scale_color_manual() and scale_fill_manual(). The result is a single plot containing not just our three smoothers but also an appropriate legend to guide the reader.

These model-fitting features make ggplot very useful for exploratory work and make it straightforward to generate and compare model-based trends and other summaries as part of the process of descriptive data visualization. The various stat_ functions are a flexible way to add summary estimates of various

Remember that we have to call two scale functions because we have two mappings, color and fill.

kinds to plots. But we will also want more than this, including presenting results from models we fit ourselves.

6.2 Look Inside Model Objects

Covering the details of fitting statistical models in R is beyond the scope of this book. For a comprehensive, modern introduction to that topic you should work your way through Gelman & Hill (2018). Harrell (2016) is also good on the many practical connections between modeling and graphing data. Similarly, Gelman (2004) provides a detailed discussion of the use of graphics as a tool in model checking and validation. Here we will discuss some ways to take the models that you fit and extract information that is easy to work with in ggplot. Our goal, as always, is to get from however the object is stored to a tidy table of numbers that we can plot. Most classes of statistical model in R will contain the information we need or will have a special set of functions, or methods, designed to extract it.

We can start by learning a little more about how the output of models is stored in R. Remember, we are always working with objects, and objects have an internal structure consisting of named pieces. Sometimes these are single numbers, sometimes vectors, and sometimes lists of things like vectors, matrices, or formulas.

We have been working extensively with tibbles and data frames. These store tables of data with named columns, perhaps consisting of different classes of variable, such as integers, characters, dates, or factors. Model objects are a little more complicated again.

```
gapminder
```

```
## # A tibble: 1,704 x 6
##    country     continent  year lifeExp       pop gdpPercap
##    <fct>       <fct>     <int>  <dbl>     <int>     <dbl>
## 1 Afghanistan Asia       1952   28.8   8425333      779.
## 2 Afghanistan Asia       1957   30.3   9240934      821.
## 3 Afghanistan Asia       1962   32.0  10267083      853.
## 4 Afghanistan Asia       1967   34.0  11537966      836.
## 5 Afghanistan Asia       1972   36.1  13079460      740.
## 6 Afghanistan Asia       1977   38.4  14880372      786.
```

```
## 7  Afghanistan Asia       1982    39.9 12881816      978.
## 8  Afghanistan Asia       1987    40.8 13867957      852.
## 9  Afghanistan Asia       1992    41.7 16317921      649.
## 10 Afghanistan Asia       1997    41.8 22227415      635.
## # ... with 1,694 more rows
```

Remember, we can use the str() function to learn more about the internal structure of any object. For example, we can get some information on what class (or classes) of object gapminder is, how large it is, and what components it has. The output from str(gapminder) is somewhat dense:

```
## Classes 'tbl_df', 'tbl' and 'data.frame':    1704 obs. of  6 variables:
## $ country : Factor w/ 142 levels "Afghanistan",..: 1 1 ...
## $ continent: Factor w/ 5 levels "Africa","Americas",..: 3 3
##    ...
## $ year : int 1952 1957 ...
## $ lifeExp : num 28.8 ...
## $ pop : int 8425333 9240934 ...
## $ gdpPercap: num 779 ...
```

There is a lot of information here about the object as a whole and each variable in it. In the same way, statistical models in R have an internal structure. But because models are more complex entities than data tables, their structure is correspondingly more complicated. There are more pieces of information, and more kinds of information, that we might want to use. All this information is generally stored in or computable from parts of a model object.

We can create a linear model, a standard OLS regression, using the gapminder data. This dataset has a country-year structure that makes an OLS specification like this the wrong one to use. But never mind that for now. We use the lm() function to run the model and store it in an object called out:

```
out ← lm(formula = lifeExp ~ gdpPercap + pop + continent, datá = gapminder)
```

The first argument is the formula for the model. lifeExp is the dependent variable and the tilde operator is used to designate the left and right sides of a model (including in cases, as we saw with facet_wrap() where the model just has a right side).

Let's look at the results by asking R to print a summary of the model.

```
summary(out)
```

```
##
## Call:
## lm(formula = lifeExp ~ gdpPercap + pop + continent, data = gapminder)
##
## Residuals:
##    Min    1Q Median    3Q    Max
## -49.16  -4.49   0.30   5.11  25.17
##
## Coefficients:
##                    Estimate Std. Error t value Pr(>|t|)
## (Intercept)        4.78e+01   3.40e-01  140.82   <2e-16 ***
## gdpPercap          4.50e-04   2.35e-05   19.16   <2e-16 ***
## pop                6.57e-09   1.98e-09    3.33    9e-04 ***
## continentAmericas  1.35e+01   6.00e-01   22.46   <2e-16 ***
## continentAsia      8.19e+00   5.71e-01   14.34   <2e-16 ***
## continentEurope    1.75e+01   6.25e-01   27.97   <2e-16 ***
## continentOceania   1.81e+01   1.78e+00   10.15   <2e-16 ***
## ---
## Signif. codes:  0 '***' 0.001 '**' 0.01 '*' 0.05 '.' 0.1 ' ' 1
##
## Residual standard error: 8.37 on 1697 degrees of freedom
## Multiple R-squared:  0.582,  Adjusted R-squared:  0.581
## F-statistic:  394 on 6 and 1697 DF,  p-value: <2e-16
```

When we use the summary() function on out, we are not getting a simple feed of what's in the model object. Instead, like any function, summary() takes its input, performs some actions, and produces output. In this case, what is printed to the console is partly information that is stored inside the model object, and partly information that the summary() function has calculated and formatted for display on the screen. Behind the scenes, summary() gets help from other functions. Objects of different classes have default *methods* associated with them, so that when the generic summary() function is applied to a linear model object, the function knows to pass the work on to a more specialized function that does a bunch of calculations

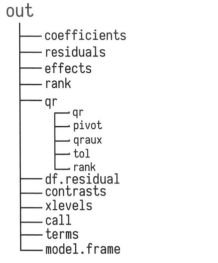

```
out
    ├── coefficients
    ├── residuals
    ├── effects
    ├── rank
    ├── qr
    │     ├── qr
    │     ├── pivot
    │     ├── qraux
    │     ├── tol
    │     └── rank
    ├── df.residual
    ├── contrasts
    ├── xlevels
    ├── call
    ├── terms
    └── model.frame
```

Figure 6.3: Schematic view of a linear model object.

Try out$df.residual at the console.

Try out$model, but be prepared for a lot of stuff to be printed at the console.

and formatting appropriate to a linear model object. We use the same generic summary() function on data frames, as in summary (gapminder), but in that case a different default method is applied.

The output from summary() gives a précis of the model, but we can't really do any further analysis with it directly. For example, what if we want to plot something from the model? The information necessary to make plots is inside the object, but it is not obvious how to use it.

If we take a look at the structure of the model object with str(out) we will find that there is a *lot* of information in there. Like most complex objects in R, out is organized as a list of components or elements. Several of these elements are themselves lists. Figure 6.3 gives you a schematic view of the contents of a linear model object. In this list of items, some elements are single values, some are data frames, and some are additional lists of simpler items. Again, remember our earlier discussion where we said objects could be thought of as being organized like a filing system: cabinets contain drawers, and a drawer may contain pages of information, whole documents, or groups of folders with more documents inside. As an alternative analogy, and sticking with the image of a list, you can think of a master to-do list for a project, where the top-level headings lead to additional lists of tasks of different kinds.

The out object created by lm contains several different named elements. Some, like the residual degrees of freedom in the model, are just a single number. Others are much larger entities, such as the data frame used to fit the model, which is retained by default. Other elements have been computed by R and then stored, such as the coefficients of the model and other quantities. You can try out$coefficients, out$residuals, and out$fitted.values, for instance. Others are lists themselves (like qr). So you can see that the summary() function is selecting and printing only a small amount of core information, in comparison to what is stored in the model object.

Just like the tables of data we saw in section 6.1, the output of summary() is presented in a way that is *compact* and *efficient* in terms of getting information across but also *untidy* when considered from the point of view of further manipulation. There is a table of coefficients, but the variable names are in the rows. The column names are awkward, and some information (e.g., at the bottom of the output) has been calculated and printed out but is not stored in the model object.

6.3 Get Model-Based Graphics Right

Figures based on statistical models face all the ordinary challenges of effective data visualization and then some. This is because model results usually carry a considerable extra burden of interpretation and necessary background knowledge. The more complex the model, the trickier it becomes to convey this information effectively, and the easier it becomes to lead one's audience or oneself into error. Within the social sciences, our ability to clearly and honestly present model-based graphics has greatly improved over the past ten or fifteen years. Over the same period, it has become clearer that some kinds of models are quite tricky to understand, even ones that had previously been seen as straightforward elements of the modeling toolkit (Ai & Norton 2003; Brambor, Clark, & Golder 2006).

Plotting model estimates is closely connected to properly estimating models in the first place. This means there is no substitute for learning the statistics. You should not use graphical methods as a substitute for understanding the model used to produce them. While this book cannot teach you that material, we can make a few general points about what good model-based graphics look like, and work through some examples of how ggplot and some additional libraries can make it easier to get good results.

Present your findings in substantive terms

Useful model-based plots show results in ways that are substantively meaningful and directly interpretable with respect to the questions the analysis is trying to answer. This means showing results in a context where other variables in the analysis are held at sensible values, such as their means or medians. With continuous variables, it can often be useful to generate predicted values that cover some substantively meaningful move across the distribution, such as from the 25th to the 75th percentile, rather than a single-unit increment in the variable of interest. For unordered categorical variables, predicted values might be presented with respect to the modal category in the data, or for a particular category of theoretical interest. Presenting substantively interpretable findings often also means using (and sometimes converting to) a scale that readers can easily understand. If your model reports results in log-odds, for example, converting the estimates to predicted probabilities will make it easier to interpret. All this advice is quite general. Each of these points applies equally well to the presentation of summary results in a table rather than a graph. There is nothing distinctively graphical about putting the focus on the substantive meaning of your findings.

Show your degree of confidence

Much the same applies to presenting the degree of uncertainty or confidence you have in your results. Model estimates come with various measures of precision, confidence, credence, or significance. Presenting and interpreting these measures is notoriously prone to misinterpretation, or overinterpretation, as researchers and audiences both demand more from things like confidence intervals and p-values than these statistics can deliver. At a minimum, having decided on an appropriate measure of model fit or the right assessment of confidence, you should show their range when you present your results. A family of ggplot geoms allows you to show a range or interval defined by position on the x-axis and then a `ymin` and `ymax` range on the y-axis. These geoms include `geom_pointrange()` and `geom_errorbar()`, which we will see in action shortly. A related geom, `geom_ribbon()` uses the same arguments to draw filled areas and is useful for plotting ranges of y-axis values along some continuously varying x-axis.

Show your data when you can

Plotting the results from a multivariate model generally means one of two things. First, we can show what is in effect a table of coefficients with associated measures of confidence, perhaps organizing the coefficients into meaningful groups, or by the size of the predicted association, or both. Second, we can show the predicted values of some variables (rather than just a model's coefficients) across some range of interest. The latter approach lets us show the original data points if we wish. The way ggplot builds graphics layer by layer allows us to easily combine model estimates (e.g., a regression line and an associated range) and the underlying data. In effect these are manually constructed versions of the automatically generated plots that we have been producing with `geom_smooth()` since the beginning of this book.

6.4 Generate Predictions to Graph

Having fitted a model, then, we might want to get a picture of the estimates it produces over the range of some particular variable, holding other covariates constant at some sensible values. The `predict()` function is a generic way of using model objects to produce this kind of prediction. In R, "generic" functions take their inputs and pass them along to more specific functions behind the scenes, ones that are suited to working with the particular kind of model object we have. The details of getting predicted values from an OLS model, for instance, will be somewhat different from getting predictions out of a logistic regression. But in each case we can use the same `predict()` function, taking care to check the documentation to see what form the results are returned in for the kind of model we are working with. Many of the most commonly used functions in R are generic in this way. The `summary()` function, for example, works on objects of many different classes, from vectors to data frames and statistical models, producing appropriate output in each case by way of a class-specific function in the background.

For `predict()` to calculate the new values for us, it needs some new data to fit the model to. We will generate a new data frame whose columns have the same names as the variables in the model's original data, but where the rows have new values. A very useful function called `expand.grid()` will help us do this. We will give it a list of variables, specifying the range of values we want each variable to take. Then `expand.grid()` will generate then will multiply out the full range of values for all combinations of the values we give it, thus creating a new data frame with the new data we need.

The function calculates the Cartesian product of the variables given to it.

In the following bit of code, we use `min()` and `max()` to get the minimum and maximum values for per capita GDP and then create a vector with one hundred evenly spaced elements between the minimum and the maximum. We hold population constant at its median, and we let continent take all its five available values.

```
min_gdp ← min(gapminder$gdpPercap)
max_gdp ← max(gapminder$gdpPercap)
med_pop ← median(gapminder$pop)

pred_df ← expand.grid(gdpPercap = (seq(from = min_gdp, to = max_gdp,
    length.out = 100)), pop = med_pop, continent = c("Africa",
    "Americas", "Asia", "Europe", "Oceania"))

dim(pred_df)
```

```
## [1] 500   3
```

```
head(pred_df)
```

```
##   gdpPercap      pop continent
## 1   241.166 7023596    Africa
## 2  1385.428 7023596    Africa
## 3  2529.690 7023596    Africa
## 4  3673.953 7023596    Africa
## 5  4818.215 7023596    Africa
## 6  5962.477 7023596    Africa
```

Now we can use predict(). If we give the function our new data and model, without any further argument it will calculate the fitted values for every row in the data frame. If we specify interval = "predict" as an argument, it will calculate 95 percent prediction intervals in addition to the point estimate.

```
pred_out ← predict(object = out, newdata = pred_df, interval = "predict")
head(pred_out)
```

```
##        fit     lwr     upr
## 1  47.9686 31.5477 64.3895
## 2  48.4830 32.0623 64.9037
## 3  48.9973 32.5767 65.4180
## 4  49.5117 33.0909 65.9325
## 5  50.0260 33.6050 66.4471
## 6  50.5404 34.1189 66.9619
```

Because we know that, by construction, the cases in `pred_df` and `pred_out` correspond row for row, we can bind the two data frames together by column. This method of joining or merging tables is *definitely not* recommended when you are dealing with data.

```
pred_df ← cbind(pred_df, pred_out)
head(pred_df)
```

```
##   gdpPercap      pop continent  fit  lwr  upr
## 1       241  7023596    Africa 48.0 31.5 64.4
## 2      1385  7023596    Africa 48.5 32.1 64.9
## 3      2530  7023596    Africa 49.0 32.6 65.4
## 4      3674  7023596    Africa 49.5 33.1 65.9
## 5      4818  7023596    Africa 50.0 33.6 66.4
## 6      5962  7023596    Africa 50.5 34.1 67.0
```

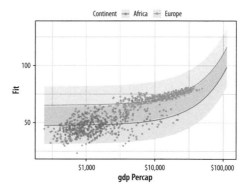

Figure 6.4: OLS predictions.

The end result is a tidy data frame, containing the predicted values from the model for the range of values we specified. Now we can plot the results. Because we produced a full range of predicted values, we can decide whether to use all of them. Here we further subset the predictions to just those for Europe and Africa (fig. 6.4).

```
p ← ggplot(data = subset(pred_df, continent %in% c("Europe", "Africa")),
        aes(x = gdpPercap,
            y = fit, ymin = lwr, ymax = upr,
            color = continent,
            fill = continent,
            group = continent))

p + geom_point(data = subset(gapminder,
                         continent %in% c("Europe", "Africa")),
            aes(x = gdpPercap, y = lifeExp,
                color = continent),
            alpha = 0.5,
            inherit.aes = FALSE) +
    geom_line() +
    geom_ribbon(alpha = 0.2, color = FALSE) +
    scale_x_log10(labels = scales::dollar)
```

We use a new geom here to draw the area covered by the prediction intervals: geom_ribbon(). It takes an x argument like a line but a ymin and ymax argument as specified in the ggplot() aesthetic mapping. This defines the lower and upper limits of the prediction interval.

In practice, you may not use predict() directly all that often. Instead, you might write code using additional packages that encapsulate the process of producing predictions and plots from models. These are especially useful when your model is a little more complex and the interpretation of coefficients becomes trickier. This happens, for instance, when you have a binary outcome variable and need to convert the results of a logistic regression into predicted probabilities, or when you have interaction terms among your predictions. We will discuss some of these helper packages in the next few sections. However, bear in mind that predict() and its ability to work safely with different classes of model underpins most of these helpers. So it's useful to see it in action firsthand in order to understand what it is doing.

6.5 Tidy Model Objects with Broom

The predict method is very useful, but there are a lot of other things we might want to do with our model output. We will use David Robinson's broom package to help us out. It is a library of functions that help us get from the model results that R generates to numbers that we can plot. It will take model objects and turn pieces of them into data frames that you can use easily with ggplot.

```
library(broom)
```

Broom takes ggplot's approach to tidy data and extends it to the model objects that R produces. Its methods can tidily extract three kinds of information. First, we can see *component-level* information about aspects of the model itself, such as coefficients and t-statistics. Second, we can obtain *observation-level* information about the model's connection to the underlying data. This includes the fitted values and residuals for each observation in the data. And finally we can get *model-level* information that summarizes the fit as a whole, such as an F-statistic, the model deviance, or the r-squared. There is a `broom` function for each of these tasks.

Get component-level statistics with tidy()

The `tidy()` function takes a model object and returns a data frame of component-level information. We can work with this to make plots in a familiar way, and much more easily than fishing inside the model object to extract the various terms. Here is an example, using the default results as just returned. For a more convenient display of the results, we will pipe the object we create with `tidy()` through a function that rounds the numeric columns of the data frame to two decimal places. This doesn't change anything about the object itself, of course.

```
out_comp ← tidy(out)
out_comp %>% round_df()
```

```
##                 term estimate std.error statistic p.value
## 1        (Intercept)    47.81      0.34    140.82       0
## 2           gdpPercap     0.00      0.00     19.16       0
## 3                 pop     0.00      0.00      3.33       0
## 4   continentAmericas    13.48      0.60     22.46       0
## 5       continentAsia     8.19      0.57     14.34       0
## 6     continentEurope    17.47      0.62     27.97       0
## 7    continentOceania    18.08      1.78     10.15       0
```

We are now able to treat this data frame just like all the other data that we have seen so far, and use it to make a plot (fig. 6.5).

```
p ← ggplot(out_comp, mapping = aes(x = term, y = estimate))

p + geom_point() + coord_flip()
```

We can extend and clean up this plot in a variety of ways. For example, we can tell `tidy()` to calculate confidence intervals for the estimates, using R's `confint()` function.

```
out_conf ← tidy(out, conf.int = TRUE)
out_conf %>% round_df()
```

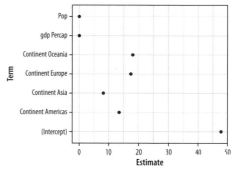

Figure 6.5: Basic plot of OLS estimates.

```
##             term estimate std.error statistic p.value conf.low conf.high
## 1    (Intercept)    47.81      0.34    140.82       0    47.15     48.48
## 2       gdpPercap     0.00      0.00     19.16       0     0.00      0.00
## 3             pop     0.00      0.00      3.33       0     0.00      0.00
## 4 continentAmericas  13.48      0.60     22.46       0    12.30     14.65
## 5    continentAsia    8.19      0.57     14.34       0     7.07      9.31
## 6  continentEurope   17.47      0.62     27.97       0    16.25     18.70
## 7 continentOceania   18.08      1.78     10.15       0    14.59     21.58
```

The convenience "not in" operator `%nin%` is available via the `socviz` library. It does the opposite of `%in%` and selects only the items in a first vector of characters that are not in the second. We'll use it to drop the intercept term from the table. We also want to do something about the labels. When fitting a model with categorical variables, R will create coefficient names based on the variable name and the category name, like `continentAmericas`. Normally we like to clean these up before plotting. Most commonly, we just want to strip away the variable name at the beginning of the coefficient label. For this we can use `prefix_strip()`, a convenience function in the `socviz` library. We tell it which prefixes to drop, using it to create a new column variable in `out_conf` that corresponds to the `terms` column but has nicer labels.

```
out_conf ← subset(out_conf, term %nin% "(Intercept)")
out_conf$nicelabs ← prefix_strip(out_conf$term, "continent")
```

Now we can use `geom_pointrange()` to make a figure (fig. 6.6) that displays some information about our confidence in the variable estimates, as opposed to just the coefficients. As with the boxplots earlier, we use `reorder()` to sort the names of the model's terms by the `estimate` variable, thus arranging our plot of effects from largest to smallest in magnitude.

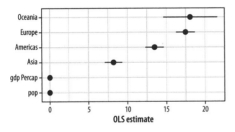

Figure 6.6: A nicer plot of OLS estimates and confidence intervals.

```
p ← ggplot(out_conf, mapping = aes(x = reorder(nicelabs, estimate),
    y = estimate, ymin = conf.low, ymax = conf.high))
p + geom_pointrange() + coord_flip() + labs(x = "", y = "OLS Estimate")
```

Dotplots of this kind can be very compact. The vertical axis can often be compressed quite a bit, with no loss in comprehension. In fact, they are often easier to read with much less room between the rows than when given a default square shape.

Get observation-level statistics with augment()

The values returned by augment() are all statistics calculated at the level of the original observations. As such, they can be added on to the data frame that the model is based on. Working from a call to augment() will return a data frame with all the original observations used in the estimation of the model, together with columns like the following:

- .fitted — the fitted values of the model
- .se.fit — the standard errors of the fitted values
- .resid — the residuals
- .hat — the diagonal of the hat matrix
- .sigma — an estimate of residual standard deviation when the corresponding observation is dropped from the model
- .cooksd — Cook's distance, a common regression diagnostic
- .std.resid — the standardized residuals

Each of these variables is named with a leading dot, for example .hat rather than hat. This is to guard against accidentally confusing it with (or accidentally overwriting) an existing variable in your data with this name. The columns of values return will differ slightly depending on the class of model being fitted.

```
out_aug ← augment(out)
head(out_aug) %>% round_df()
```

```
##    lifeExp gdpPercap       pop continent .fitted .se.fit .resid .hat .sigma .cooksd .std.resid
## 1     28.8       779   8425333      Asia    56.4    0.47  -27.6    0   8.34    0.01      -3.31
## 2     30.3       821   9240934      Asia    56.4    0.47  -26.1    0   8.34    0.00      -3.13
## 3     32.0       853  10267083      Asia    56.5    0.47  -24.5    0   8.35    0.00      -2.93
## 4     34.0       836  11537966      Asia    56.5    0.47  -22.4    0   8.35    0.00      -2.69
## 5     36.1       740  13079460      Asia    56.4    0.47  -20.3    0   8.35    0.00      -2.44
## 6     38.4       786  14880372      Asia    56.5    0.47  -18.0    0   8.36    0.00      -2.16
```

By default, augment() will extract the available data from the model object. This will usually include the variables used in the model itself but not any additional ones contained in the original data frame. Sometimes it is useful to have these. We can add them by specifying the data argument:

```
out_aug ← augment(out, data = gapminder)
head(out_aug) %>% round_df()
```

##		country	continent	year	lifeExp	pop	gdpPercap	.fitted	.se.fit	.resid	.hat	.sigma	.cooksd
##	1	Afghanistan	Asia	1952	28.8	8425333	779	56.4	0.47	-27.6	0	8.34	0.01
##	2	Afghanistan	Asia	1957	30.3	9240934	821	56.4	0.47	-26.1	0	8.34	0.00
##	3	Afghanistan	Asia	1962	32.0	10267083	853	56.5	0.47	-24.5	0	8.35	0.00
##	4	Afghanistan	Asia	1967	34.0	11537966	836	56.5	0.47	-22.4	0	8.35	0.00
##	5	Afghanistan	Asia	1972	36.1	13079460	740	56.4	0.47	-20.3	0	8.35	0.00
##	6	Afghanistan	Asia	1977	38.4	14880372	786	56.5	0.47	-18.0	0	8.36	0.00

##		.std.resid
##	1	-3.31
##	2	-3.13
##	3	-2.93
##	4	-2.69
##	5	-2.44
##	6	-2.16

If some rows containing missing data were dropped to fit the model, then these will not be carried over to the augmented data frame.

The new columns created by augment() can be used to create some standard regression plots. For example, we can plot the residuals versus the fitted values. Figure 6.7 suggests, unsurprisingly, that our country-year data has rather more structure than is captured by our OLS model.

```
p ← ggplot(data = out_aug, mapping = aes(x = .fitted, y = .resid))
p + geom_point()
```

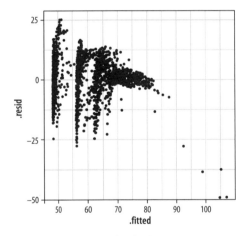

Figure 6.7: Residuals vs fitted values.

Get model-level statistics with glance()

The glance() function organizes the information typically presented at the bottom of a model's summary() output. By itself, it usually just returns a table with a single row in it. But as we shall see in a moment, the real power of broom's approach is the way that it can scale up to cases where we are grouping or subsampling our data.

```
glance(out) %>% round_df()
```

```
##    r.squared adj.r.squared sigma statistic p.value df
## 1      0.58            0.58  8.37    393.91       0  7
##      logLik     AIC     BIC deviance df.residual
## 1 -6033.83 12083.6 12127.2   118754        1697
```

Broom is able to tidy (and augment, and glance at) a wide range of model types. Not all functions are available for all classes of model. Consult broom's documentation for more details on what is available. For example, here is a plot created from the tidied output of an event-history analysis. First we generate a Cox proportional hazards model of some survival data.

```
library(survival)

out_cph ← coxph(Surv(time, status) ~ age + sex, data = lung)
out_surv ← survfit(out_cph)
```

The details of the fit are not important here, but in the first step the Surv() function creates the response or outcome variable for the proportional hazards model that is then fitted by the coxph() function. Then the survfit() function creates the survival curve from the model, much like we used predict() to generate predicted values earlier. Try summary(out_cph) to see the model, and summary(out_surv) to see the table of predicted values that will form the basis for our plot. Next we tidy out_surv to get a data frame, and plot it (fig. 6.8).

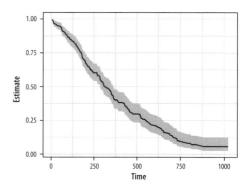

Figure 6.8: A Kaplan-Meier plot.

```
out_tidy ← tidy(out_surv)

p ← ggplot(data = out_tidy, mapping = aes(time, estimate))
p + geom_line() + geom_ribbon(mapping = aes(ymin = conf.low,
    ymax = conf.high), alpha = 0.2)
```

6.6 Grouped Analysis and List Columns

Broom makes it possible to quickly fit models to different subsets of your data and get consistent and usable tables of results out the other end. Let's say we wanted to look at the gapminder data by examining the relationship between life expectancy and GDP by *continent*, for each year in the data. The gapminder data is at bottom organized by country-years. That is the unit of observation in the rows. If we wanted, we could take a slice of the data manually, such as "all countries observed in Asia, in 1962" or "all in Africa, 2002." Here is "Europe, 1977":

```
eu77 ← gapminder %>% filter(continent == "Europe", year == 1977)
```

We could then see what the relationship between life expectancy and GDP looked like for that continent-year group:

```
fit ← lm(lifeExp ~ log(gdpPercap), data = eu77)
summary(fit)
```

```
##
## Call:
## lm(formula = lifeExp ~ log(gdpPercap), data = eu77)
##
## Residuals:
##    Min     1Q Median     3Q    Max
## -7.496 -1.031  0.093  1.176  3.712
##
## Coefficients:
##                 Estimate Std. Error t value Pr(>|t|)
## (Intercept)       29.489      7.161    4.12  0.00031 ***
## log(gdpPercap)     4.488      0.756    5.94  2.2e-06 ***
## ---
## Signif. codes:
## 0 '***' 0.001 '**' 0.01 '*' 0.05 '.' 0.1 ' ' 1
##
## Residual standard error: 2.11 on 28 degrees of freedom
## Multiple R-squared:  0.557,  Adjusted R-squared:  0.541
## F-statistic: 35.2 on 1 and 28 DF,  p-value: 2.17e-06
```

With dplyr and broom we can do this for every continent-year slice of the data in a compact and tidy way. We start with our table of data and then (%>%) group the countries by continent and year using the group_by() function. We introduced this grouping operation in chapter 4. Our data is reorganized first by continent, and within continent by year. Here we will take one further step and *nest* the data that makes up each group:

```
out_le ← gapminder %>%
    group_by(continent, year) %>%
    nest()

out_le
```

```
## # A tibble: 60 x 3
##    continent  year data
##    <fct>     <int> <list>
##  1 Asia       1952 <tibble [33 x 4]>
##  2 Asia       1957 <tibble [33 x 4]>
##  3 Asia       1962 <tibble [33 x 4]>
##  4 Asia       1967 <tibble [33 x 4]>
##  5 Asia       1972 <tibble [33 x 4]>
##  6 Asia       1977 <tibble [33 x 4]>
##  7 Asia       1982 <tibble [33 x 4]>
##  8 Asia       1987 <tibble [33 x 4]>
##  9 Asia       1992 <tibble [33 x 4]>
## 10 Asia       1997 <tibble [33 x 4]>
## # ... with 50 more rows
```

Think of what nest() does as a more intensive version what group_by() does. The resulting object is has the tabular form we expect (it is a tibble), but it looks a little unusual. The first two columns are the familiar continent and year. But we now also have a new column, data, that contains a small table of data corresponding to each continent-year group. This is a *list column*, something we have not seen before. It turns out to be very useful for bundling together complex objects (structured, in this case, as a list of tibbles, each being a 33 x 4 table of data) within the rows of our data (which remains tabular). Our "Europe 1977" fit is in there. We can look at it, if we like, by filtering the data and then *unnesting* the list column.

```
out_le %>% filter(continent == "Europe" & year == 1977) %>% unnest()
```

```
## # A tibble: 30 x 6
##    continent  year country  lifeExp    pop gdpPercap
##    <fct>     <int> <fct>      <dbl>  <int>     <dbl>
##  1 Europe     1977 Albania     68.9 2.51e6     3533.
##  2 Europe     1977 Austria     72.2 7.57e6    19749.
```

```
##  3 Europe     1977 Belgium          72.8 9.82e6    19118.
##  4 Europe     1977 Bosnia and Her~  69.9 4.09e6     3528.
##  5 Europe     1977 Bulgaria         70.8 8.80e6     7612.
##  6 Europe     1977 Croatia          70.6 4.32e6    11305.
##  7 Europe     1977 Czech Republic   70.7 1.02e7    14800.
##  8 Europe     1977 Denmark          74.7 5.09e6    20423.
##  9 Europe     1977 Finland          72.5 4.74e6    15605.
## 10 Europe     1977 France           73.8 5.32e7    18293.
## # ... with 20 more rows
```

List columns are useful because we can act on them in a compact and tidy way. In particular, we can pass functions along to each row of the list column and make something happen. For example, a moment ago we ran a regression of life expectancy and logged GDP for European countries in 1977. We can do that for every continent-year combination in the data. We first create a convenience function called fit_ols() that takes a single argument, df (for data frame), and that fits the linear model we are interested in. Then we *map* that function to each of our list column rows in turn. Recall from chapter 4 that mutate creates new variables or columns within a pipeline.

The map action is an important idea in functional programming. If you have written code in other, more imperative languages, you can think of it as a compact alternative to writing for … next loops. You can of course write loops like this in R. Computationally they are often not any less efficient than their functional alternatives. But mapping functions to arrays is more easily integrated into a sequence of data transformations.

```r
fit_ols ← function(df) {
    lm(lifeExp ~ log(gdpPercap), data = df)
}


out_le ← gapminder %>%
    group_by(continent, year) %>%
    nest() %>%
    mutate(model = map(data, fit_ols))


out_le
```

```
## # A tibble: 60 x 4
##    continent year data              model
##    <fct>     <int> <list>            <list>
##  1 Asia       1952 <tibble [33 x 4]> <S3: lm>
##  2 Asia       1957 <tibble [33 x 4]> <S3: lm>
##  3 Asia       1962 <tibble [33 x 4]> <S3: lm>
##  4 Asia       1967 <tibble [33 x 4]> <S3: lm>
##  5 Asia       1972 <tibble [33 x 4]> <S3: lm>
```

```
##  6 Asia      1977 <tibble [33 x 4]> <S3: lm>
##  7 Asia      1982 <tibble [33 x 4]> <S3: lm>
##  8 Asia      1987 <tibble [33 x 4]> <S3: lm>
##  9 Asia      1992 <tibble [33 x 4]> <S3: lm>
## 10 Asia      1997 <tibble [33 x 4]> <S3: lm>
## # ... with 50 more rows
```

Before starting the pipeline we create a new function: it is a convenience function whose only job is to estimate a particular OLS model on some data. Like almost everything in R, functions are a kind of object. To make a new one, we use the slightly special function() function. There is a little more detail on creating functions in the appendix. To see what fit_ols() looks like once it is created, type fit_ols without parentheses at the console. To see what it does, try fit_ols(df = gapminder) or summary(fit_ols(gapminder)).

Now we have two list columns: data and model. The latter was created by mapping the fit_ols() function to each row of data. Inside each element of model is a linear model for that continent-year. So we now have sixty OLS fits, one for every continent-year grouping. Having the models inside the list column is not much use to us in and of itself. But we can extract the information we want while keeping things in a tidy tabular form. For clarity we will run the pipeline from the beginning again, this time adding a few new steps.

First we extract summary statistics from each model by mapping the tidy() function from broom to the model list column. Then we unnest the result, dropping the other columns in the process. Finally, we filter out all the Intercept terms and also drop all observations from Oceania. In the case of the Intercepts we do this just out of convenience. Oceania we drop just because there are so few observations. We put the results in an object called out_tidy.

```
fit_ols ← function(df) {
    lm(lifeExp ~ log(gdpPercap), data = df)
}


out_tidy ← gapminder %>%
    group_by(continent, year) %>%
    nest() %>%
```

```
    mutate(model = map(data, fit_ols),
           tidied = map(model, tidy)) %>%
    unnest(tidied, .drop = TRUE) %>%
    filter(term %nin% "(Intercept)" &
           continent %nin% "Oceania")


out_tidy %>% sample_n(5)
```

```
## # A tibble: 5 x 7
##   continent  year term          estimate std.error statistic   p.value
##   <fct>     <int> <chr>            <dbl>     <dbl>     <dbl>      <dbl>
## 1 Europe     1987 log(gdpPercap)    4.14     0.752      5.51 0.00000693
## 2 Asia       1972 log(gdpPercap)    4.44     1.01       4.41 0.000116
## 3 Europe     1972 log(gdpPercap)    4.51     0.757      5.95 0.00000208
## 4 Americas   1952 log(gdpPercap)   10.4      2.72       3.84 0.000827
## 5 Asia       1987 log(gdpPercap)    5.17     0.727      7.12 0.0000000531
```

We now have tidy regression output with an estimate of the
association between log GDP per capita and life expectancy for
each year, within continents. We can plot these estimates (fig. 6.9)
in a way that takes advantage of their groupiness.

```
p ← ggplot(data = out_tidy,
           mapping = aes(x = year, y = estimate,
                         ymin = estimate - 2*std.error,
                         ymax = estimate + 2*std.error,
                         group = continent, color = continent))
```

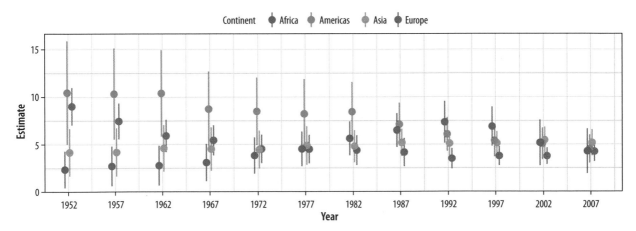

Figure 6.9: Yearly estimates of the association between GDP and life expectancy, pooled by continent.

```
p + geom_pointrange(position = position_dodge(width = 1)) +
    scale_x_continuous(breaks = unique(gapminder$year)) +
    theme(legend.position = "top") +
    labs(x = "Year", y = "Estimate", color = "Continent")
```

The call to `position_dodge()` within `geom_pointrange()` allows the point ranges for each continent to be near one another within years, instead of being plotted right on top of one another. We could have faceted the results by continent, but doing it this way lets us see differences in the yearly estimates much more easily. This technique is very useful not just for cases like this but also when you want to compare the coefficients given by different kinds of statistical model. This sometimes happens when we're interested in seeing how, say, OLS performs against some other model specification.

6.7 Plot Marginal Effects

Our earlier discussion of `predict()` was about obtaining estimates of the average effect of some coefficient, net of the other terms in the model. Over the past decade, estimating and plotting *partial* or *marginal effects* from a model has become an increasingly common way of presenting accurate and interpretively useful predictions. Interest in marginal effects plots was stimulated by the realization that the interpretation of terms in logistic regression models, in particular, was trickier than it seemed—especially when there were interaction terms in the model (Ai & Norton 2003). Thomas Leeper's `margins` package can make these plots for us.

```
library(margins)
```

To see it in action, we'll take another look at the General Social Survey data in `gss_sm`, this time focusing on the binary variable `obama`. It is coded 1 if the respondent said he or she voted for Barack Obama in the 2012 presidential election and 0 otherwise. In this case, mostly for convenience, the zero code includes all other answers to the question, including those who said they voted for Mitt Romney, those who said they did not vote, those who refused to answer, and those who said they didn't know who

As is common with retrospective questions on elections, rather more people claim to have voted for Obama than is consistent with the vote share he received in the election.

they voted for. We will fit a logistic regression on obama, with age, polviews, race, and sex as the predictors. The age variable is the respondent's age in years. The sex variable is coded as "Male" or "Female," with "Male" as the reference category. The race variable is coded as "White," "Black," or "Other," with "White" as the reference category. The polviews measure is a self-reported scale of the respondent's political orientation from "Extremely Conservative" through "Extremely Liberal," with "Moderate" in the middle. We take polviews and create a new variable, polviews_m, using the relevel() function to recode "Moderate" to be the reference category. We fit the model with the glm() function and specify an interaction between race and sex.

```
gss_sm$polviews_m ← relevel(gss_sm$polviews, ref = "Moderate")

out_bo ← glm(obama ~ polviews_m + sex*race,
             family = "binomial", data = gss_sm)
summary(out_bo)
```

```
##
## Call:
## glm(formula = obama ~ polviews_m + sex * race, family = "binomial",
##     data = gss_sm)
##
## Deviance Residuals:
##    Min     1Q  Median     3Q    Max
## -2.905  -0.554   0.177   0.542  2.244
##
## Coefficients:
##                                    Estimate Std. Error z value Pr(>|z|)
## (Intercept)                         0.29649    0.13409    2.21   0.0270 *
## polviews_mExtremely Liberal         2.37295    0.52504    4.52  6.2e-06 ***
## polviews_mLiberal                   2.60003    0.35667    7.29  3.1e-13 ***
## polviews_mSlightly Liberal          1.29317    0.24843    5.21  1.9e-07 ***
## polviews_mSlightly Conservative    -1.35528    0.18129   -7.48  7.7e-14 ***
## polviews_mConservative             -2.34746    0.20038  -11.71  < 2e-16 ***
## polviews_mExtremely Conservative   -2.72738    0.38721   -7.04  1.9e-12 ***
## sexFemale                           0.25487    0.14537    1.75   0.0796 .
## raceBlack                           3.84953    0.50132    7.68  1.6e-14 ***
## raceOther                          -0.00214    0.43576    0.00   0.9961
## sexFemale:raceBlack                -0.19751    0.66007   -0.30   0.7648
```

```
## sexFemale:raceOther                 1.57483    0.58766    2.68   0.0074 **
## ---
## Signif. codes:  0 '***' 0.001 '**' 0.01 '*' 0.05 '.' 0.1 ' ' 1
##
## (Dispersion parameter for binomial family taken to be 1)
##
##     Null deviance: 2247.9  on 1697  degrees of freedom
## Residual deviance: 1345.9  on 1686  degrees of freedom
##   (1169 observations deleted due to missingness)
## AIC: 1370
##
## Number of Fisher Scoring iterations: 6
```

The summary reports the coefficients and other information. We can now graph the data in any one of several ways. Using margins(), we calculate the marginal effects for each variable:

```
bo_m ← margins(out_bo)
summary(bo_m)
```

```
##                               factor     AME      SE        z        p   lower   upper
##            polviews_mConservative  -0.4119  0.0283  -14.5394  0.0000  -0.4674  -0.3564
##  polviews_mExtremely Conservative  -0.4538  0.0420  -10.7971  0.0000  -0.5361  -0.3714
##       polviews_mExtremely Liberal   0.2681  0.0295    9.0996  0.0000   0.2103   0.3258
##                 polviews_mLiberal   0.2768  0.0229   12.0736  0.0000   0.2319   0.3218
##  polviews_mSlightly Conservative  -0.2658  0.0330   -8.0596  0.0000  -0.3304  -0.2011
##       polviews_mSlightly Liberal   0.1933  0.0303    6.3896  0.0000   0.1340   0.2526
##                         raceBlack   0.4032  0.0173   23.3568  0.0000   0.3694   0.4371
##                         raceOther   0.1247  0.0386    3.2297  0.0012   0.0490   0.2005
##                         sexFemale   0.0443  0.0177    2.5073  0.0122   0.0097   0.0789
```

The margins library comes with several plot methods of its own. If you wish, at this point you can just try plot(bo_m) to see a plot of the average marginal effects, produced with the general look of a Stata graphic. Other plot methods in the margins library include cplot(), which visualizes marginal effects conditional on a second variable, and image(), which shows predictions or marginal effects as a filled heatmap or contour plot.

Alternatively, we can take results from margins() and plot them ourselves. To clean up the summary a little, we convert it to a tibble, then use prefix_strip() and prefix_replace() to

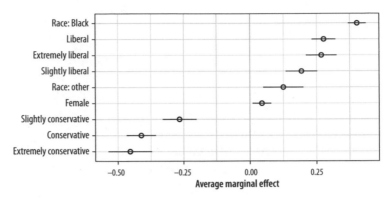

tidy the labels. We want to strip the polviews_m and sex prefixes and (to avoid ambiguity about "Other") adjust the race prefix.

```
bo_gg ← as_tibble(summary(bo_m))
prefixes ← c("polviews_m", "sex")
bo_gg$factor ← prefix_strip(bo_gg$factor, prefixes)
bo_gg$factor ← prefix_replace(bo_gg$factor, "race", "Race: ")


bo_gg %>% select(factor, AME, lower, upper)
```

```
## # A tibble: 9 x 4
##    factor                     AME    lower    upper
## *  <chr>                    <dbl>    <dbl>    <dbl>
## 1  Conservative            -0.412  -0.467   -0.356
## 2  Extremely Conservative  -0.454  -0.536   -0.371
## 3  Extremely Liberal        0.268   0.210    0.326
## 4  Liberal                  0.277   0.232    0.322
## 5  Slightly Conservative   -0.266  -0.330   -0.201
## 6  Slightly Liberal         0.193   0.134    0.253
## 7  Race: Black              0.403   0.369    0.437
## 8  Race: Other              0.125   0.0490   0.200
## 9  Female                   0.0443  0.00967  0.0789
```

Now we have a table that we can plot (fig. 6.10) as we have learned:

```
p ← ggplot(data = bo_gg, aes(x = reorder(factor, AME),
                             y = AME, ymin = lower, ymax = upper))


p + geom_hline(yintercept = 0, color = "gray80") +
```

```
geom_pointrange() + coord_flip() +
labs(x = NULL, y = "Average Marginal Effect")
```

If we are just interested in getting conditional effects for a particular variable, then conveniently we can ask the plot methods in the margins library to do the work calculating effects for us but without drawing their plot. Instead, they can return the results in a format we can easily use in ggplot, and with less need for cleanup. For example, with cplot() we can draw figure 6.11.

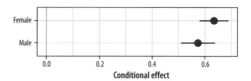

Figure 6.11: Conditional effects plot.

```
pv_cp ← cplot(out_bo, x = "sex", draw = FALSE)

p ← ggplot(data = pv_cp, aes(x = reorder(xvals, yvals),
                             y = yvals, ymin = lower, ymax = upper))

p + geom_hline(yintercept = 0, color = "gray80") +
    geom_pointrange() + coord_flip() +
    labs(x = NULL, y = "Conditional Effect")
```

The margins package is under active development. It can do much more than described here. The vignettes that come with the package provide more extensive discussion and numerous examples.

6.8 Plots from Complex Surveys

Social scientists often work with data collected using a complex survey design. Survey instruments may be stratified by region or some other characteristic, contain replicate weights to make them comparable to a reference population, have a clustered structure, and so on. In chapter 4 we learned how calculate and then plot frequency tables of categorical variables, using some data from the General Social Survey. However, if we want accurate estimates of U.S. households from the GSS, we will need to take the survey's design into account and use the survey weights provided in the dataset. Thomas Lumley's survey library provides a comprehensive set of tools for addressing these issues. The tools and the theory behind them are discussed in detail in Lumley

(2010), and an overview of the package is provided in Lumley (2004). While the functions in the survey package are straightforward to use and return results in a generally tidy form, the package predates the tidyverse and its conventions by several years. This means we cannot use survey functions directly with dplyr. However, Greg Freedman Ellis has written a helper package, srvyr, that solves this problem for us and lets us use the survey library's functions within a data analysis pipeline in a familiar way.

For example, the gss_lon data contains a small subset of measures from every wave of the GSS since its inception in 1972. It also contains several variables that describe the design of the survey and provide replicate weights for observations in various years. These technical details are described in the GSS documentation. Similar information is typically provided by other complex surveys. Here we will use this design information to calculate weighted estimates of the distribution of educational attainment by race for selected survey years from 1976 to 2016.

To begin, we load the survey and srvyr libraries.

```
library(survey)
library(srvyr)
```

Next, we take our gss_lon dataset and use the survey tools to create a new object that contains the data, as before, but with some additional information about the survey's design:

```
options(survey.lonely.psu = "adjust")
options(na.action="na.pass")

gss_wt ← subset(gss_lon, year > 1974) %>%
    mutate(stratvar = interaction(year, vstrat)) %>%
    as_survey_design(ids = vpsu,
                     strata = stratvar,
                     weights = wtssall,
                     nest = TRUE)
```

The two options set at the beginning provides some information to the survey library about how to behave. You should consult Lumley (2010) and the survey package documentation for details. The subsequent operations create gss_wt, an object with one additional column (stratvar), describing the yearly sampling strata. We use the interaction() function to do this. It multiplies the vstrat variable by the year variable to get a vector of stratum information for each year. In the next step, we use the as_survey_design() function to add the key pieces of information about the survey design. It adds information about the sampling identifiers (ids), the strata (strata), and the replicate weights (weights). With those in place we can take advantage of a large number of specialized functions in the survey library that allow us to calculate properly weighted survey means or estimate models with the correct sampling specification. For example, we can calculate the distribution of education by race for a series of years from 1976 to 2016. We use survey_mean() to do this:

We have to do this because of the way the GSS codes its stratum information.

```
out_grp ← gss_wt %>%
    filter(year %in% seq(1976, 2016, by = 4)) %>%
    group_by(year, race, degree) %>%
    summarize(prop = survey_mean(na.rm = TRUE))

out_grp
```

```
## # A tibble: 150 x 5
##     year race  degree              prop prop_se
##    <dbl> <fct> <fct>              <dbl>   <dbl>
##  1 1976. White Lt High School    0.328   0.0160
##  2 1976. White High School       0.518   0.0162
##  3 1976. White Junior College   0.0129  0.00298
##  4 1976. White Bachelor          0.101   0.00960
##  5 1976. White Graduate         0.0393   0.00644
##  6 1976. Black Lt High School    0.562   0.0611
##  7 1976. Black High School       0.337   0.0476
##  8 1976. Black Junior College   0.0426   0.0193
##  9 1976. Black Bachelor         0.0581    0.0239
## 10 1976. Black Graduate          0.       0.
## # ... with 140 more rows
```

The results returned in out_grp include standard errors. We can also ask survey_mean() to calculate confidence intervals for us, if we wish.

Grouping with group_by() lets us calculate counts or means for the innermost variable, grouped by the next variable "up" or "out," in this case, degree by race, such that the proportions for degree will sum to one for each group in race, and this will be done separately for each value of year. If we want the *marginal* frequencies, such that the values for all combinations of race and degree sum to one within each year, we first have to interact the variables we are cross-classifying. Then we group by the new interacted variable and do the calculation as before:

```
out_mrg ← gss_wt %>%
    filter(year %in% seq(1976, 2016, by = 4)) %>%
    mutate(racedeg = interaction(race, degree)) %>%
    group_by(year, racedeg) %>%
    summarize(prop = survey_mean(na.rm = TRUE))

out_mrg
```

```
## # A tibble: 150 x 4
##     year racedeg                prop prop_se
##    <dbl> <fct>                 <dbl>   <dbl>
##  1 1976. White.Lt High School 0.298  0.0146
##  2 1976. Black.Lt High School 0.0471 0.00840
##  3 1976. Other.Lt High School 0.00195 0.00138
##  4 1976. White.High School    0.471  0.0160
##  5 1976. Black.High School    0.0283 0.00594
##  6 1976. Other.High School    0.00325 0.00166
##  7 1976. White.Junior College 0.0117 0.00268
##  8 1976. Black.Junior College 0.00357 0.00162
##  9 1976. Other.Junior College 0.      0.
## 10 1976. White.Bachelor       0.0919 0.00888
## # ... with 140 more rows
```

This gives us the numbers that we want and returns them in a tidy data frame. The interaction() function produces variable labels that are a compound of the two variables we interacted, with each combination of categories separated by a period (such as White.Graduate). However, perhaps we would like to see these categories as two separate columns, one for race and one for education, as before. Because the variable labels are organized in a predictable way, we can use one of the convenient functions in the tidyverse's `tidyr` package to separate the single variable into two columns while correctly preserving the row values. Appropriately, this function is called `separate()`.

```
out_mrg ← gss_wt %>% filter(year %in% seq(1976, 2016, by = 4)) %>%
    mutate(racedeg = interaction(race, degree)) %>% group_by(year,
    racedeg) %>% summarize(prop = survey_mean(na.rm = TRUE)) %>%
    separate(racedeg, sep = "\\.", into = c("race", "degree"))

out_mrg
```

The two backslashes before the period in the call to **separate** are necessary for R to interpret it literally as a period. By default in search-and-replace operations like this, the search terms are regular expressions. The period acts as a special character, a kind of wildcard, meaning "any character at all." To make the regular expression engine treat it literally, we add one backslash before it. The backslash is an "escape" character. It means "The next character is going to be treated differently from usual." However, because the backslash is a special character as well, we need to add a second backslash to make sure the parser sees it properly.

```
## # A tibble: 150 x 5
##     year race  degree              prop  prop_se
##    <dbl> <chr> <chr>              <dbl>    <dbl>
##  1 1976. White Lt High School    0.298   0.0146
##  2 1976. Black Lt High School    0.0471  0.00840
##  3 1976. Other Lt High School    0.00195 0.00138
##  4 1976. White High School       0.471   0.0160
##  5 1976. Black High School       0.0283  0.00594
##  6 1976. Other High School       0.00325 0.00166
##  7 1976. White Junior College    0.0117  0.00268
##  8 1976. Black Junior College    0.00357 0.00162
##  9 1976. Other Junior College    0.      0.
## 10 1976. White Bachelor          0.0919  0.00888
## # ... with 140 more rows
```

The call to `separate()` says to take the `racedeg` column, split each value when it sees a period, and reorganize the results into two columns, `race` and `degree`. This gives us a tidy table much like `out_grp` but for the marginal frequencies.

Reasonable people can disagree over how best to plot a small multiple of a frequency table while faceting by year, especially when there is some measure of uncertainty attached. A barplot is the obvious approach for a single case, but when there are many years it can be difficult to compare bars across panels. This is especially the case when standard errors or confidence intervals are used in conjunction with bars. This is sometimes called a "dynamite plot," not because it looks amazing but because the t-shaped error bars on the tops of the columns make them look like cartoon dynamite plungers. An alternative is to use a line graph to join up the time observations, faceting on educational categories

Sometimes it may be preferable to show that the underlying variable is categorical, as a bar chart makes clear, and not continuous, as a line graph suggests. Here the trade-off is in favor of the line graphs as the bars are hard to compare across facets.

Figure 6.12: Weighted estimates of educational attainment for whites and blacks, GSS selected years 1976–2016. Faceting barplots is often a bad idea, and the more facets there are, the worse an idea it is. With a small-multiple plot the viewer wants to compare across panels (in this case, over time), but this is difficult to do when the data inside the panels are categorical comparisons shown as bars (in this case, education level by group).

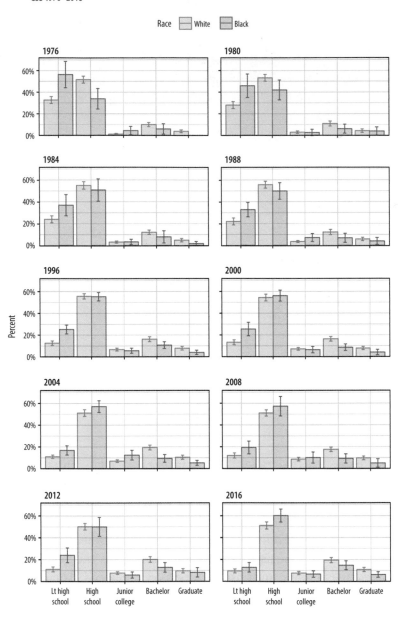

Educational attainment by race

GSS 1976–2016

instead of year. Figure 6.12 shows the results for our GSS data in dynamite-plot form, where the error bars are defined as twice the standard error in either direction around the point estimate.

```r
p ← ggplot(data = subset(out_grp, race %nin% "Other"),
           mapping = aes(x = degree, y = prop,
                         ymin = prop - 2*prop_se,
                         ymax = prop + 2*prop_se,
                         fill = race,
                         color = race,
                         group = race))

dodge ← position_dodge(width=0.9)

p + geom_col(position = dodge, alpha = 0.2) +
    geom_errorbar(position = dodge, width = 0.2) +
    scale_x_discrete(labels = scales::wrap_format(10)) +
    scale_y_continuous(labels = scales::percent) +
    scale_color_brewer(type = "qual", palette = "Dark2") +
    scale_fill_brewer(type = "qual", palette = "Dark2") +
    labs(title = "Educational Attainment by Race",
         subtitle = "GSS 1976-2016",
         fill = "Race",
         color = "Race",
         x = NULL, y = "Percent") +
    facet_wrap(~ year, ncol = 2) +
    theme(legend.position = "top")
```

This plot has a few cosmetic details and adjustments that we will learn more about in chapter 8. As before, I encourage you to peel back the plot from the bottom, one instruction at a time, to see what changes. One useful adjustment to notice is the new call to the scales library to adjust the labels on the x-axis. The adjustment on the y-axis is familiar, scales::percent to convert the proportion to a percentage. On the x-axis, the issue is that several of the labels are rather long. If we do not adjust them they will print over one another. The scales::wrap_format() function will break long labels into lines. It takes a single numerical argument (here 10) that is the maximum length a string can be before it is wrapped onto a new line.

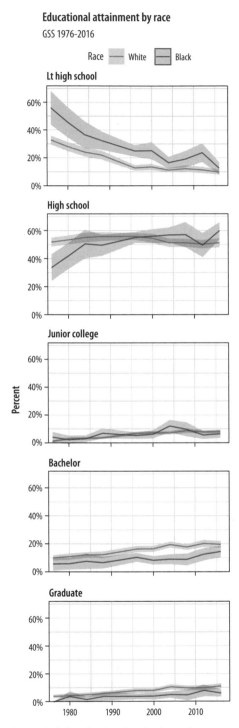

Educational attainment by race
GSS 1976-2016

Figure 6.13: Faceting by education instead.

A graph like this is true to the categorical nature of the data, while showing the breakdown of groups within each year. But you should experiment with some alternatives. For example, we might decide that it is better to facet by degree category instead, and put the year on the x-axis within each panel. If we do that, then we can use geom_line() to show a time trend, which is more natural, and geom_ribbon() to show the error range. This is perhaps a better way to show the data, especially as it brings out the time trends within each degree category and allows us to see the similarities and differences by racial classification at the same time (fig. 6.13).

```
p ← ggplot(data = subset(out_grp, race %in% "Other"),
           mapping = aes(x = year, y = prop, ymin = prop - 2*prop_se,
                         ymax = prop + 2*prop_se, fill = race, color = race,
                         group = race))


p + geom_ribbon(alpha = 0.3, aes(color = NULL)) +
    geom_line() +
    facet_wrap(~ degree, ncol = 1) +
    scale_y_continuous(labels = scales::percent) +
    scale_color_brewer(type = "qual", palette = "Dark2") +
    scale_fill_brewer(type = "qual", palette = "Dark2") +
    labs(title = "Educational Attainment by Race",
         subtitle = "GSS 1976-2016", fill = "Race",
         color = "Race", x = NULL, y = "Percent") +
    theme(legend.position = "top")
```

6.9 Where to Go Next

In general, when you estimate models and want to plot the results, the difficult step is not the plotting but rather calculating and extracting the right numbers. Generating predicted values and measures of confidence or uncertainty from models requires that you understand the model you are fitting and the function you use to fit it, especially when it involves interactions, cross-level effects, or transformations of the predictor or response scales. The details can vary substantially from model type to model type, and also with the goals of any particular analysis. It is unwise to approach them mechanically. That said, several tools exist to help you work with model objects and produce a default set of plots from them.

Default plots for models

Just as model objects in R usually have a default `summary()` method, printing out an overview tailored to the type of model it is, they will usually have a default `plot()` method, too. Figures produced by `plot()` are typically not generated via ggplot, but it is usually worth exploring them. They typically make use of either R's base graphics or the `lattice` library (Sarkar 2008). These are two plotting systems not covered in this book. Default plot methods are easy to examine. Let's take a look again at our simple OLS model.

```
out ← lm(formula = lifeExp ~ log(gdpPercap) + pop + continent,
    data = gapminder)
```

To look at some of R's default plots for this model, use the `plot()` function.

```
# Plot not shown
plot(out, which = c(1, 2), ask = FALSE)
```

The `which()` statement here selects the first two of four default plots for this kind of model. If you want to easily reproduce base R's default model graphics using ggplot, the `ggfortify` package is worth examining. It is similar to `broom` in that it tidies the output of model objects, but it focuses on producing a standard plot (or group of plots) for a wide variety of model types. It does this by defining a function called `autoplot()`. The idea is to be able to use `autoplot()` with the output of many different kinds of model.

A second option worth looking at is the `coefplot` package. It provides a quick way to produce good-quality plots of point estimates and confidence intervals (fig. 6.14). It has the advantage of managing the estimation of interaction effects and other occasionally tricky calculations.

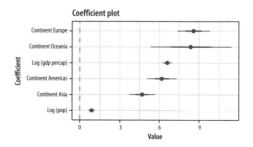

Figure 6.14: A plot from coefplot.

```
library(coefplot)
out ← lm(formula = lifeExp ~ log(gdpPercap) + log(pop) + continent,
    data = gapminder)


coefplot(out, sort = "magnitude", intercept = FALSE)
```

Tools in development

Tidyverse tools for modeling and model exploration are being actively developed. The broom and margins packages continue to get more and more useful. There are also other projects worth paying attention to. The infer package is in its early stages but can already do useful things in a pipeline-friendly way. You can install it from CRAN with install.packages("infer").

infer.netlify.com

Extensions to ggplot

The GGally package provides a suite of functions designed to make producing standard but somewhat complex plots a little easier. For instance, it can produce generalized pairs plots, a useful way of quickly examining possible relationships between several different variables at once. This sort of plot is like the visual version of a correlation matrix. It shows a bivariate plot for all pairs of variables in the data. This is relatively straightforward when all the variables are continuous measures. Things get more complex when, as is often the case in the social sciences, some or all variables are categorical or otherwise limited in the range of values they can take. A generalized pairs plot can handle these cases. For example, figure 6.15 shows a generalized pairs plot for five variables from the organdata dataset.

```
library(GGally)


organdata_sm ← organdata %>% select(donors, pop_dens, pubhealth,
    roads, consent_law)


ggpairs(data = organdata_sm, mapping = aes(color = consent_law),
    upper = list(continuous = wrap("density"), combo = "box_no_facet"),
    lower = list(continuous = wrap("points"), combo = wrap("dot_no_facet")))
```

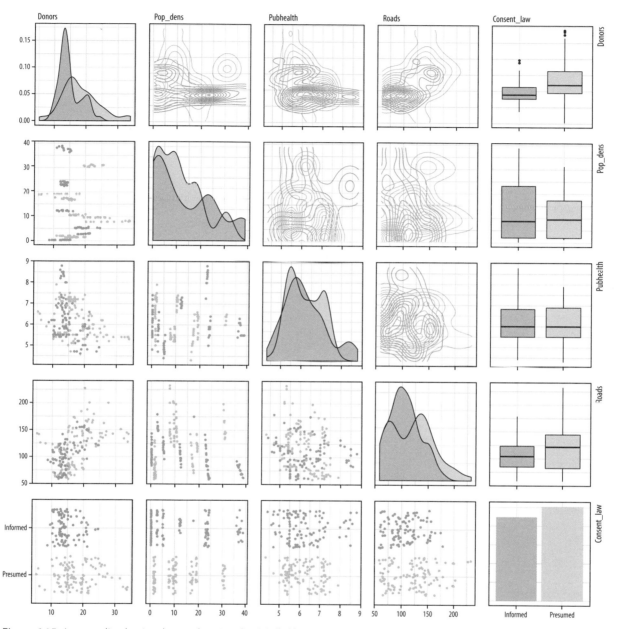

Figure 6.15: A generalized pairs plot made using the GGally library.

Multipanel plots like those in figure 6.15 are intrinsically very rich in information. When combined with several within-panel types of representation, or any more than a modest number of variables, they can become quite complex. They should be used sparingly for the presentation of finished work. More often they are a useful tool for a researcher to quickly investigate aspects of a data set. The goal is not to pithily summarize a single point one already knows but to open things up for further exploration.

7   Draw Maps

Choropleth maps show geographical regions colored, shaded, or graded according to some variable. They are visually striking, especially when the spatial units of the map are familiar entities, like the countries of the European Union, or states in the United States. But maps like this can also sometimes be misleading. Although it is not a dedicated Geographical Information System (GIS), R can work with geographical data, and ggplot can make choropleth maps. But we'll also consider some other ways of representing data of this kind.

Figure 7.1 shows a series of maps of the U.S. presidential election results in 2016. Reading from the top left, we see, first, a state-level map where the margin of victory can be high (a darker blue or red) or low (a lighter blue or red). The color scheme has no midpoint. Second, we see a county-level map colored bright red or blue depending on the winner. Third is a county-level map where the color of red and blue counties is graded by the size of the vote share. Again, the color scale has no midpoint. Fourth is a county-level map with a continuous color gradient from blue to red, but that passes through a purple midpoint for areas where the balance of the vote is close to even. The map in the bottom left has the same blue-purple-red scheme but distorts the geographical boundaries by squeezing or inflating them to reflect the population of the county shown. Finally in the bottom right we see a cartogram, where states are drawn using square tiles, and the number of tiles each state gets is proportional to the number of electoral college votes it has (which in turn is proportional to that state's population).

Each of these maps shows data for the same event, but the impressions they convey are very different. Each faces two main problems. First, the underlying quantities of interest are only partly spatial. The number of electoral college votes won and the share of votes cast within a state or county are expressed in spatial terms, but ultimately it is the number of people within those regions that matter. Second, the regions themselves are of wildly

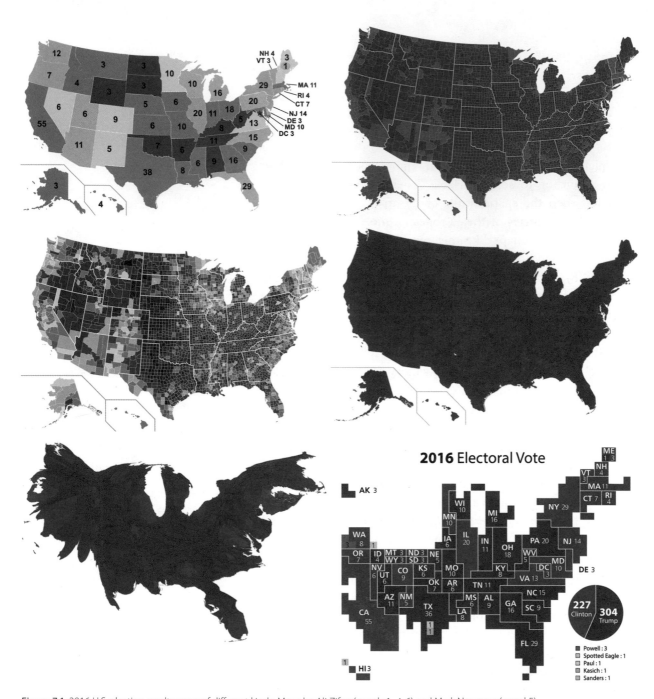

Figure 7.1: 2016 U.S. election results maps of different kinds. Maps by Ali Zifan (panels 1–4, 6) and Mark Newman (panel 5).

differing sizes, and they differ in a way that is not well correlated with the magnitudes of the underlying votes. The mapmakers also face choices that would arise in many other representations of the data. Do we want to just show who won each state in absolute terms (this is all that matters for the actual result, in the end) or do we want to indicate how close the race was? Do we want to display the results at some finer level of resolution than is relevant to the outcome, such as county rather than state counts? How can we convey that different data points can carry very different weights because they represent vastly larger or smaller numbers of people? It is tricky enough to convey these choices honestly with different colors and shape sizes on a simple scatterplot. Often a map is like a weird grid that you are forced to conform to even though you know it systematically misrepresents what you want to show.

This is not always the case, of course. Sometimes our data really is purely spatial, and we can observe it at a fine enough level of detail that we can represent spatial distributions honestly and in a compelling way. But the spatial features of much social science are collected through entities such as precincts, neighborhoods, metro areas, census tracts, counties, states, and nations. These may themselves be socially contingent. A great deal of cartographic work with social-scientific variables involves working both with and against that arbitrariness. Geographers call this the Modifiable Areal Unit Problem, or MAUP (Openshaw 1983).

7.1 Map U.S. State-Level Data

Let's take a look at some data for the U.S. presidential election in 2016 and see how we might plot it in R. The election dataset has various measures of the vote and vote shares by state. Here we pick some columns and sample a few rows at random.

```
election %>% select(state, total_vote,
                    r_points, pct_trump, party, census) %>%
    sample_n(5)
```

```
## # A tibble: 5 x 6
##   state       total_vote r_points pct_trump party      census
##   <chr>            <dbl>    <dbl>     <dbl> <chr>      <chr>
## 1 Kentucky       1924149.     29.8      62.5 Republican South
```

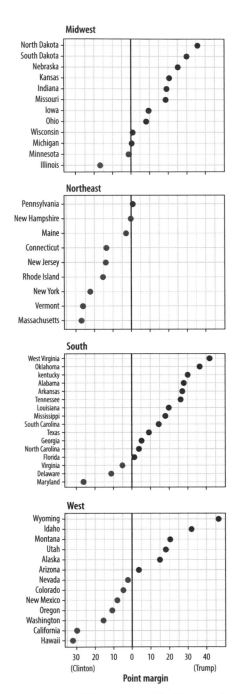

Figure 7.2: 2016 election results. Would a two-color choropleth map be more informative than this, or less?

## 2	Vermont	315067.	-26.4	30.3 Democrat	Northeast
## 3	South Carolina	2103027.	14.3	54.9 Republican	South
## 4	Wyoming	255849.	46.3	68.2 Republican	West
## 5	Kansas	1194755.	20.4	56.2 Republican	Midwest

The FIPS code is a federal code that numbers states and territories of the United States. It extends to the county level with an additional four digits, so every U.S. county has a unique six-digit identifier, where the first two digits represent the state. This dataset also contains the census region of each state.

```
# Hex color codes for Dem Blue and Rep Red
party_colors ← c("#2E74C0", "#CB454A")


p0 ← ggplot(data = subset(election, st %nin% "DC"),
         mapping = aes(x = r_points,
                       y = reorder(state, r_points),
                       color = party))


p1 ← p0 + geom_vline(xintercept = 0, color = "gray30") +
    geom_point(size = 2)


p2 ← p1 + scale_color_manual(values = party_colors)


p3 ← p2 + scale_x_continuous(breaks = c(-30, -20, -10, 0, 10, 20, 30, 40),
                    labels = c("30\n (Clinton)", "20", "10", "0",
                               "10", "20", "30", "40\n(Trump)"))


p3 + facet_wrap(~ census, ncol=1, scales="free_y") +
    guides(color=FALSE) + labs(x = "Point Margin", y = "") +
    theme(axis.text=element_text(size=8))
```

The first thing you should remember about spatial data is that you don't have to represent it spatially. We've been working with country-level data throughout and have yet to make a map of it. Of course, spatial representations can be very useful, and sometimes absolutely necesssary. But we can start with a state-level dotplot, faceted by region (fig. 7.2). This figure brings together many aspects of plot construction that we have worked on so far, including subsetting data, reordering results by a second variable, and using a scale formatter. It also introduces some new options, like manually setting the color of an aesthetic. We break up the construction process into several steps by creating intermediate objects (p0, p1, p2) along the way. This makes the code more readable. Bear in mind also that, as always, you can try plotting each of these intermediate objects as well (just type their name at the console and hit return) to see what they look like. What happens if you remove the scales="free_y" argument to facet_wrap()? What happens if you delete the call to scale_color_manual()?

As always, the first task in drawing a map is to get a data frame with the right information in it, and in the right order. First we load R's maps package, which provides us with some predrawn map data.

```
library(maps)
us_states ← map_data("state")
head(us_states)
```

```
##      long     lat group order  region subregion
## 1 -87.4620 30.3897     1     1 alabama      <NA>
## 2 -87.4849 30.3725     1     2 alabama      <NA>
## 3 -87.5250 30.3725     1     3 alabama      <NA>
## 4 -87.5308 30.3324     1     4 alabama      <NA>
## 5 -87.5709 30.3267     1     5 alabama      <NA>
## 6 -87.5881 30.3267     1     6 alabama      <NA>
```

```
dim(us_states)
```

```
## [1] 15537     6
```

This just a data frame. It has more than 15,000 rows because you need a lot of lines to draw a good-looking map. We can make a blank state map right away with this data, using geom_polygon().

```
p ← ggplot(data = us_states, mapping = aes(x = long, y = lat,
    group = group))

p + geom_polygon(fill = "white", color = "black")
```

The map in figure 7.3 is plotted with latitude and longitude points, which are there as scale elements mapped to the x- and y-axes. A map is, after all, just a set of lines drawn in the right order on a grid.

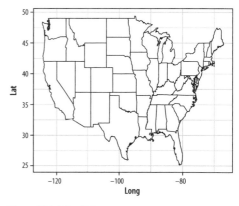

Figure 7.3: A first U.S. map.

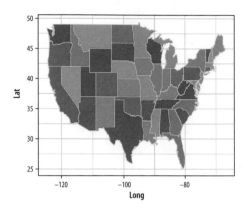

Figure 7.4: Coloring the states.

We can map the fill aesthetic to `region` and change the `color` mapping to a light gray and thin the lines to make the state borders a little nicer (fig. 7.4). We'll also tell R not to plot a legend.

```
p ← ggplot(data = us_states, aes(x = long, y = lat, group = group,
    fill = region))

p + geom_polygon(color = "gray90", size = 0.1) + guides(fill = FALSE)
```

Next, let's deal with the projection. By default the map is plotted using the venerable Mercator projection. It doesn't look that good. Assuming we are not planning on sailing across the Atlantic, the practical virtues of this projection are not much use to us, either. If you glance again at the maps in figure 7.1, you'll notice they look nicer. This is because they are using an Albers projection. (Look, for example, at the way that the U.S.-Canadian border is a little curved along the 49th parallel from Washington state to Minnesota, rather than a straight line.) Techniques for map projection are a fascinating world of their own, but for now just remember we can transform the default projection used by geom_polygon() via the coord_map() function. Remember that we said that projection onto a coordinate system is a necessary part of the plotting process for any data. Normally it is left implicit. We have not usually had to specify a coord_ function because most of the time we have drawn our plots on a simple Cartesian plane. Maps are more complex. Our locations and borders are defined on a more or less spherical object, which means we must have a method for transforming or projecting our points and lines from a round to a flat surface. The many ways of doing this give us a menu of cartographic options.

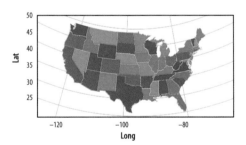

Figure 7.5: Improving the projection.

The Albers projection requires two latitude parameters, `lat0` and `lat1`. We give them their conventional values for a U.S. map here (fig. 7.5). (Try messing around with their values and see what happens when you redraw the map.)

```
p ← ggplot(data = us_states,
            mapping = aes(x = long, y = lat,
                        group = group, fill = region))

p + geom_polygon(color = "gray90", size = 0.1) +
    coord_map(projection = "albers", lat0 = 39, lat1 = 45) +
    guides(fill = FALSE)
```

Now we need to get our own data onto the map. Remember, underneath that map is just a big data frame specifying a large number of lines that need to be drawn. We have to merge our data with that data frame. Somewhat annoyingly, in the map data the state names (in a variable named `region`) are in lowercase. We can create a variable in our own data frame to correspond to this, using the `tolower()` function to convert the `state` names. We then use `left_join` to merge, but you could also use `merge(..., sort = FALSE)`. This merge step is important! You need to take care that the values of the key variables you are matching on really do exactly correspond to one another. If they do not, missing values (`NA` codes) will be introduced into your merge, and the lines on your map will not join up. This will result in a weirdly "sharded" appearance to your map when R tries to fill the polygons. Here, the `region` variable is the only column with the same name in both the data sets we are joining, and so the `left_join()` function uses it by default. If the keys have different names in each data set, you can specify that if needed.

To reiterate, it is important to know your data and variables well enough to check that they have merged properly. Do not do it blindly. For example, if rows corresponding to Washington, DC, were named "washington dc" in the `region` variable of your `election` data frame but "district of columbia" in the corresponding `region` variable of your map data, then merging on `region` would mean no rows in the `election` data frame would match "washington dc" in the map data, and the resulting merged variables for those rows would all be coded as missing. Maps that look broken when you draw them are usually caused by merge errors. But errors can also be subtle. For example, perhaps one of your state names inadvertently has a leading (or, worse, a trailing) space as a result of the data originally being imported from elsewhere and not fully cleaned. That would mean, for example, that `california` and `california ⌴` are different strings, and the match would fail. In ordinary use you might not easily see the extra space (designated here by ⌴). So be careful.

```
election$region ← tolower(election$state)
us_states_elec ← left_join(us_states, election)
```

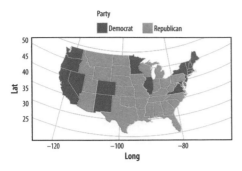

Figure 7.6: Mapping the results.

Figure 7.7: Election 2016 by state.

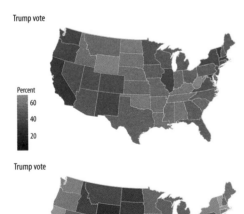

Figure 7.8: Two versions of percent Trump by state.

We have now merged the data. Take a look at the object with head(us_states_elec). Now that everything is in one big data frame, we can plot it on a map (fig. 7.6).

```
p ← ggplot(data = us_states_elec,
            aes(x = long, y = lat,
                group = group, fill = party))


p + geom_polygon(color = "gray90", size = 0.1) +
    coord_map(projection = "albers", lat0 = 39, lat1 = 45)
```

To complete the map (fig. 7.7), we will use our party colors for the fill, move the legend to the bottom, and add a title. Finally we will remove the grid lines and axis labels, which aren't really needed, by defining a special theme for maps that removes most of the elements we don't need. (We'll learn more about themes in chapter 8. You can also see the code for the map theme in the appendix.)

```
p0 ← ggplot(data = us_states_elec,
            mapping = aes(x = long, y = lat,
                          group = group, fill = party))
p1 ← p0 + geom_polygon(color = "gray90", size = 0.1) +
    coord_map(projection = "albers", lat0 = 39, lat1 = 45)
p2 ← p1 + scale_fill_manual(values = party_colors) +
    labs(title = "Election Results 2016", fill = NULL)
p2 + theme_map()
```

With the map data frame in place, we can map other variables if we like. Let's try a continuous measure, such as the percentage of the vote received by Donald Trump. To begin with, in figure 7.8 we just map the variable we want (pct_trump) to the fill aesthetic and see what geom_polygon() does by default.

```
p0 ← ggplot(data = us_states_elec,
            mapping = aes(x = long, y = lat, group = group, fill = pct_trump))


p1 ← p0 + geom_polygon(color = "gray90", size = 0.1) +
    coord_map(projection = "albers", lat0 = 39, lat1 = 45)


p1 + labs(title = "Trump vote") + theme_map() + labs(fill = "Percent")
```

```
p2 ← p1 + scale_fill_gradient(low = "white", high = "#CB454A") +
        labs(title = "Trump vote")
p2 + theme_map() + labs(fill = "Percent")
```

The default color used in the p1 object is blue. Just for reasons of convention, that isn't what is wanted here. In addition, the gradient runs in the wrong direction. In our case, the standard interpretation is that a higher vote share makes for a darker color. We fix both of these problems in the p2 object by specifying the scale directly. We'll use the values we created earlier in party_colors.

For election results, we might prefer a gradient that diverges from a midpoint. The scale_gradient2() function gives us a blue-red spectrum that passes through white by default. Alternatively, we can respecify the midlevel color along with the high and low colors. We will make purple our midpoint and use the muted() function from the scales library to tone down the color a little.

```
p0 ← ggplot(data = us_states_elec,
              mapping = aes(x = long, y = lat, group = group, fill = d_points))

p1 ← p0 + geom_polygon(color = "gray90", size = 0.1) +
    coord_map(projection = "albers", lat0 = 39, lat1 = 45)

p2 ← p1 + scale_fill_gradient2() + labs(title = "Winning margins")
p2 + theme_map() + labs(fill = "Percent")

p3 ← p1 + scale_fill_gradient2(low = "red", mid = scales::muted("purple"),
                                high = "blue", breaks = c(-25, 0, 25, 50, 75)) +
    labs(title = "Winning margins")
p3 + theme_map() + labs(fill = "Percent")
```

If you look at the gradient scale for this first "purple America" map, in figure 7.9, you'll see that it extends very high on the blue side. This is because Washington, DC, is included in the data and hence the scale. Even though it is barely visible on the map, DC has by far the highest points margin in favor of the Democrats of any unit of observation in the data. If we omit it, we'll see that our scale shifts in a way that does not just affect the top of the blue end but recenters the whole gradient and makes the red side more vivid as a result. Figure 7.10 shows the result.

Winning margins

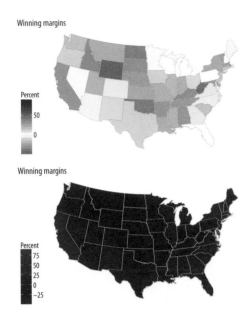

Percent

50

0

Winning margins

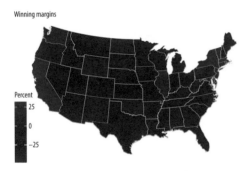

Percent
75
50
25
0
−25

Figure 7.9: Two views of Trump vs Clinton share: a white midpoint and a purple America version.

Winning margins

Percent
25
0
−25

Figure 7.10: A purple America version of Trump vs Clinton that excludes results from Washington, DC.

```
p0 ← ggplot(data = subset(us_states_elec,
                          region %nin% "district of columbia"),
            aes(x = long, y = lat, group = group, fill = d_points))

p1 ← p0 + geom_polygon(color = "gray90", size = 0.1) +
    coord_map(projection = "albers", lat0 = 39, lat1 = 45)

p2 ← p1 + scale_fill_gradient2(low = "red",
                               mid = scales::muted("purple"),
                               high = "blue") +
    labs(title = "Winning margins")
p2 + theme_map() + labs(fill = "Percent")
```

This brings out the familiar choropleth problem of having geographical areas that only partially represent the variable we are mapping. In this case, we're showing votes spatially, but what really matters is the number of people who voted in each state.

7.2 America's Ur-choropleths

In the U.S. case, administrative areas vary widely in geographical area and also in population size. The modifiable areal unit problem evident at the state level, as we have seen, also arises even more at the county level. County-level U.S. maps can be aesthetically pleasing because of the added detail they bring to a national map. But they also make it easy to present a geographical distribution to insinuate an explanation. The results can be tricky to work with. When producing county maps, it is important to remember that the states of New Hampshire, Rhode Island, Massachussetts, and Connecticut are all smaller in area than any of the ten largest Western *counties*. Many of those counties have fewer than a hundred thousand people living in them. Some have fewer than ten thousand inhabitants.

The result is that most choropleth maps of the United States for whatever variable in effect show population density more than anything else. The other big variable, in the U.S. case, is percent black. Let's see how to draw these two maps in R. The procedure is essentially the same as it was for the state-level map. We need two data frames, one containing the map data, and the other containing the fill variables we want plotted. Because there are more than three thousand counties in the United States, these two data frames will be rather larger than they were for the state-level maps.

The datasets are included in the socviz library. The county map data frame has already been processed a little in order to transform it to an Albers projection, and also to relocate (and rescale) Alaska and Hawaii so that they fit into an area in the bottom left of the figure. This is better than throwing away two states from the data. The steps for this transformation and relocation are not shown here. If you want to see how it's done, consult the appendix for details. Let's take a look at our county map data first:

```
county_map %>% sample_n(5)
```

```
##              long      lat  order  hole piece              group     id
## 116977    -286097 -1302531 116977 FALSE     1 0500000US35025.1  35025
## 175994    1657614  -698592 175994 FALSE     1 0500000US51197.1  51197
## 186409     674547   -65321 186409 FALSE     1 0500000US55011.1  55011
## 22624      619876 -1093164  22624 FALSE     1 0500000US05105.1  05105
## 5906     -1983421 -2424955   5906 FALSE    10 0500000US02016.10 02016
```

It looks the same as our state map data frame, but it is much larger, running to almost 200,000 rows. The id field is the FIPS code for the county. Next, we have a data frame with county-level demographic, geographic, and election data:

```
county_data %>%
    select(id, name, state, pop_dens, pct_black) %>%
    sample_n(5)
```

```
##          id              name state    pop_dens      pct_black
## 3029  53051 Pend Oreille County    WA [   0,   10) [ 0.0, 2.0)
## 1851  35041    Roosevelt County    NM [   0,   10) [ 2.0, 5.0)
## 1593  29165       Platte County    MO [ 100,  500) [ 5.0,10.0)
## 2363  45009      Bamberg County    SC [  10,   50) [50.0,85.3]
## 654   17087     Johnson County    IL [  10,   50) [ 5.0,10.0)
```

This data frame includes information for entities besides counties, though not for all variables. If you look at the top of the object with head(), you'll notice that the first row has an id of 0. Zero is the FIPS code for the entire United States, and thus the data in this row are for the whole country. Similarly, the second row has an id of 01000, which corresponds to the state FIPS of 01, for the whole of Alabama. As we merge county_data in to county_map,

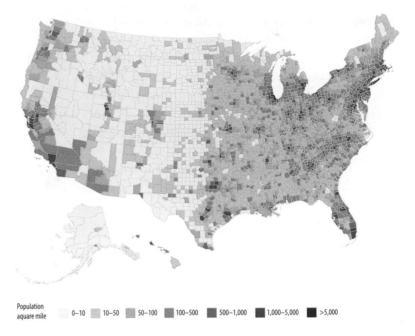

<table>
<tr><td>Population
aquare mile</td><td>0–10</td><td>10–50</td><td>50–100</td><td>100–500</td><td>500–1,000</td><td>1,000–5,000</td><td>>5,000</td></tr>
</table>

these state rows will be dropped, along with the national row, as county_map only has county-level data.

We merge the data frames using the shared FIPS id column:

```
county_full ← left_join(county_map, county_data, by = "id")
```

With the data merged, we can map the population density per square mile (fig. 7.11).

```
p ← ggplot(data = county_full,
           mapping = aes(x = long, y = lat,
                         fill = pop_dens,
                         group = group))


p1 ← p + geom_polygon(color = "gray90", size = 0.05) + coord_equal()


p2 ← p1 + scale_fill_brewer(palette="Blues",
                            labels = c("0-10", "10-50", "50-100", "100-500",
                                       "500-1,000", "1,000-5,000", ">5,000"))


p2 + labs(fill = "Population per\nsquare mile") +
    theme_map() +
    guides(fill = guide_legend(nrow = 1)) +
    theme(legend.position = "bottom")
```

If you try out the p1 object, you will see that ggplot produces a legible map, but by default it chooses an unordered categorical layout. This is because the pop_dens variable is not ordered. We could recode it so that R is aware of the ordering. Alternatively, we can manually supply the right sort of scale using the scale_fill_brewer() function, together with a nicer set of labels. We will learn more about this scale function in the next chapter. We also tweak how the legend is drawn using the guides() function to make sure each element of the key appears on the same row. Again, we will see this use of guides() in more detail in the next chapter. The use of coord_equal() makes sure that the relative scale of our map does not change even if we alter the overall dimensions of the plot.

We can now do exactly the same thing for our map of percent black population by county (fig. 7.12). Once again, we specify a palette for the fill mapping using scale_fill_brewer(), this time choosing a different range of hues for the map.

```
p ← ggplot(data = county_full,
           mapping = aes(x = long, y = lat, fill = pct_black,
                         group = group))
p1 ← p + geom_polygon(color = "gray90", size = 0.05) + coord_equal()
p2 ← p1 + scale_fill_brewer(palette="Greens")
```

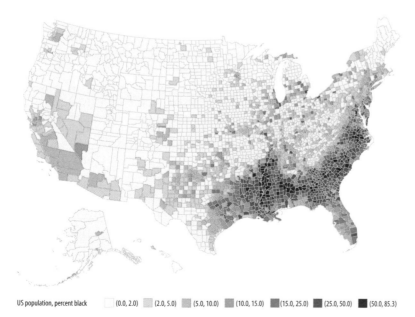

Figure 7.12: Percent black population by county.

US population, percent black   (0.0, 2.0) (2.0, 5.0) (5.0, 10.0) (10.0, 15.0) (15.0, 25.0) (25.0, 50.0) (50.0, 85.3)

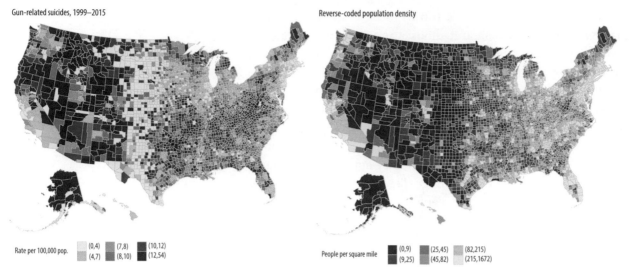

Gun-related suicides, 1999–2015

Reverse-coded population density

Rate per 100,000 pop.   (0,4) (7,8) (10,12)
                         (4,7) (8,10) (12,54)

People per square mile   (0,9) (25,45) (82,215)
                          (9,25) (45,82) (215,1672)

Figure 7.13: Gun-related suicides by county; reverse-coded population density by county. Before tweeting this picture, please read the text for discussion of what is wrong with it.

```
p2 + labs(fill = "US Population, Percent Black") +
    guides(fill = guide_legend(nrow = 1)) +
    theme_map() + theme(legend.position = "bottom")
```

Figures 7.11 and 7.12 are America's "ur-choropleths." Between the two of them, population density and percent black will do a lot to obliterate many a suggestively patterned map of the United States. These two variables aren't *explanations* of anything in isolation, but if it turns out that it is more useful to know one or both of them instead of the thing you're plotting, you probably want to reconsider your theory.

As an example of the problem in action, let's draw two new county-level choropleths (fig. 7.13). The first is an effort to replicate a poorly sourced but widely circulated county map of firearm-related suicide rates in the United States. The su_gun6 variable in county_data (and county_full) is a measure of the rate of all firearm-related suicides between 1999 and 2015. The rates are binned into six categories. We have a pop_dens6 variable that divides the population density into six categories, too.

We first draw a map with the su_gun6 variable. We will match the color palettes between the maps, but for the population map we will flip our color scale around so that less populated areas are shown in a darker shade. We do this by using a function from the RColorBrewer library to manually create two palettes. The rev() function used here reverses the order of a vector.

```
orange_pal ← RColorBrewer::brewer.pal(n = 6, name = "Oranges")
orange_pal
```

```
## [1] "#FEEDDE" "#FDD0A2" "#FDAE6B" "#FD8D3C" "#E6550D"
## [6] "#A63603"
```

```
orange_rev ← rev(orange_pal)
orange_rev
```

```
## [1] "#A63603" "#E6550D" "#FD8D3C" "#FDAE6B" "#FDD0A2"
## [6] "#FEEDDE"
```

The brewer.pal() function produces evenly spaced color schemes to order from any one of several named palettes. The colors are specified in hexadecimal format. Again, we will learn more about color specifications and how to manipulate palettes for mapped variables in chapter 8.

```
gun_p ← ggplot(data = county_full,
            mapping = aes(x = long, y = lat,
                        fill = su_gun6,
                        group = group))

gun_p1 ← gun_p + geom_polygon(color = "gray90", size = 0.05) + coord_equal()

gun_p2 ← gun_p1 + scale_fill_manual(values = orange_pal)

gun_p2 + labs(title = "Gun-Related Suicides, 1999-2015",
            fill = "Rate per 100,000 pop.") +
    theme_map() + theme(legend.position = "bottom")
```

Having drawn the gun plot, we use almost exactly the same code to draw the reverse-coded population density map.

```
pop_p ← ggplot(data = county_full, mapping = aes(x = long, y = lat,
                                    fill = pop_dens6,
                                    group = group))
```

```
pop_p1 ← pop_p + geom_polygon(color = "gray90", size = 0.05) + coord_equal()


pop_p2 ← pop_p1 + scale_fill_manual(values = orange_rev)


pop_p2 + labs(title = "Reverse-coded Population Density",
              fill = "People per square mile") +
   theme_map() + theme(legend.position = "bottom")
```

It's clear that the two maps are not identical. However, the visual impact of the first has a lot in common with the second. The dark bands in the West (except for California) stand out, and they fade as we move toward the center of the country. There are some strong similarities elsewhere on the map too, such as in the Northeast.

The gun-related suicide measure is already expressed as a rate. It is the number of qualifying deaths in a county, divided by that county's population. Normally, we standardize in this way to "control for" the fact that larger populations will tend to produce more gun-related suicides just because they have more people in them. However, this sort of standardization has its limits. In particular, when the event of interest is not very common, and there is very wide variation in the base size of the units, then the denominator (e.g., the population size) starts to be expressed more and more in the standardized measure.

Third, and more subtly, the data is subject to reporting constraints connected to population size. If there are fewer than ten events per year for a cause of death, the Centers for Disease Control (CDC) will not report them at the county level because it might be possible to identify particular deceased individuals. Assigning data like this to bins creates a threshold problem for choropleth maps. Look again figure 7.13. The gun-related suicides panel seems to show a north-south band of counties with the lowest rate of suicides running from the Dakotas down through Nebraska, Kansas, and into West Texas. Oddly, this band borders counties in the West with the very highest rates, from New Mexico on up. But from the density map we can see that many counties in both these regions have very low population densities. Are they really that different in their gun-related suicide rates?

Probably not. More likely, we are seeing an artifact arising from how the data is coded. For example, imagine a county with

100,000 inhabitants that experiences nine gun-related suicides in a year. The CDC will not report this number. Instead it will be coded as "suppressed," accompanied by a note saying any standardized estimates or rates will also be unreliable. But if we are determined to make a map where all the counties are colored in, we might be tempted to put any suppressed results into the lowest bin. After all, we know that the number is somewhere between zero and ten. Why not just code it as zero? Meanwhile, a county with 100,000 inhabitants that experiences twelve gun-related suicides a year *will* be numerically reported. The CDC is a responsible organization, and so although it provides the absolute number of deaths for all counties above the threshold, the notes to the data file will still warn you that any rate calculated with this number will be unreliable. If we do it anyway, then twelve deaths in a small population might well put a sparsely populated county in the highest category of suicide rate. Meanwhile, a low-population county just under that threshold would be coded as being in the lowest (lightest) bin. But in reality they might not be so different, and in any case efforts to quantify that difference will be unreliable. If estimates for these counties cannot be obtained directly or estimated with a good model, then it is better to drop those cases as missing, even at the cost of your beautiful map, than have large areas of the country painted with a color derived from an unreliable number.

Small differences in reporting, combined with coarse binning and miscoding, will produce spatially misleading and substantively mistaken results. It might seem that focusing on the details of variable coding in this particular case is a little too much in the weeds for a general introduction. But it is exactly these details that can dramatically alter the appearance of any graph, and especially maps, in a way that can be hard to detect after the fact.

Do not do this. One standard alternative is to estimate the suppressed observations using a count model. An approach like this might naturally lead to more extensive, properly spatial modeling of the data.

7.3 Statebins

As an alternative to state-level choropleths, we can consider *statebins*, using a package developed by Bob Rudis. We will use it to look again at our state-level election results. Statebins is similar to ggplot but has a slightly different syntax from the one we're used to. It needs several arguments, including the basic

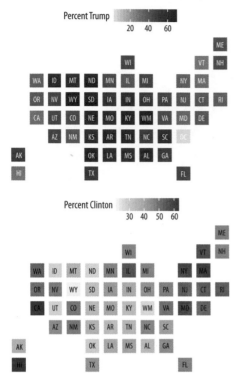

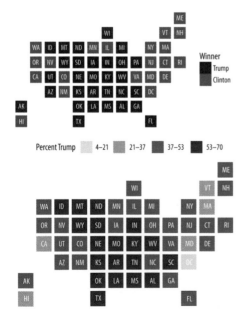

Figure 7.14: Statebins of the election results. We omit DC from the Clinton map to prevent the scale becoming unbalanced.

Figure 7.15: Manually specifying colors for statebins.

data frame (the `state_data` argument), a vector of state names (`state_col`), and the value being shown (`value_col`). In addition, we can optionally tell it the color palette we want to use and the color of the text to label the state boxes. For a continuous variable we can use `statebins_continuous()`, as follows, to make figure 7.14:

```
library(statebins)

statebins_continuous(state_data = election, state_col = "state",
                     text_color = "white", value_col = "pct_trump",
                     brewer_pal="Reds", font_size = 3,
                     legend_title="Percent Trump")

statebins_continuous(state_data = subset(election, st %nin% "DC"),
                     state_col = "state",
                     text_color = "black", value_col = "pct_clinton",
                     brewer_pal="Blues", font_size = 3,
                     legend_title="Percent Clinton")
```

Sometimes we will want to present categorical data. If our variable is already cut into categories we can use `statebins_manual()` to represent it. Here we add a new variable to the `election` data called `color`, just mirroring party names with two appropriate color names. We do this because we need to specify the colors we are using by way of a variable in the data frame, not as a proper mapping. We tell the `statebins_manual()` function that the colors are contained in a column named `color` and use it for the first map in figure 7.15.

Alternatively, we can have `statebins()` cut the data using the `breaks` argument, as in the second plot in figure 7.15.

```
election ← election %>% mutate(color = recode(party, Republican = "darkred",
                                              Democrat = "royalblue"))

statebins_manual(state_data = election, state_col = "st",
                 color_col = "color", text_color = "white",
                 font_size = 3, legend_title="Winner",
                 labels=c("Trump", "Clinton"), legend_position = "right")
```

```
statebins(state_data = election,
        state_col = "state", value_col = "pct_trump",
        text_color = "white", breaks = 4,
        labels = c("4-21", "21-37", "37-53", "53-70"),
        brewer_pal="Reds", font_size = 3, legend_title="Percent Trump")
```

7.4 Small-Multiple Maps

Sometimes we have geographical data with repeated observations over time. A common case is to have a country- or state-level measure observed over a period of years. In these cases, we might want to make a small-multiple map to show changes over time. For example, the opiates data has state-level measures of the death rate from opiate-related causes (such as heroin or fentanyl overdoses) between 1999 and 2014.

```
opiates
```

```
## # A tibble: 800 x 11
##    year state       fips deaths population crude adjusted
##    <int> <chr>      <int> <int>     <int> <dbl>   <dbl>
## 1  1999 Alabama        1    37   4430141 0.800   0.800
## 2  1999 Alaska         2    27    624779 4.30    4.00
## 3  1999 Arizona        4   229   5023823 4.60    4.70
## 4  1999 Arkansas       5    28   2651860 1.10    1.10
## 5  1999 California     6  1474  33499204 4.40    4.50
## 6  1999 Colorado       8   164   4226018 3.90    3.70
## 7  1999 Connecticut    9   151   3386401 4.50    4.40
## 8  1999 Delaware      10    32    774990 4.10    4.10
## 9  1999 District o~   11    28    570213 4.90    4.90
## 10 1999 Florida       12   402  15759421 2.60    2.60
## # ... with 790 more rows, and 4 more variables:
## #   adjusted_se <dbl>, region <ord>, abbr <chr>,
## #   division_name <chr>
```

As before, we can take our us_states object, the one with the state-level map details, and merge it with our opiates dataset. As before, we convert the state variable in the opiates data to lowercase first, to make the match work properly.

```
opiates$region ← tolower(opiates$state)
opiates_map ← left_join(us_states, opiates)
```

Because the `opiates` data includes the `year` variable, we are now in a position to make a faceted small-multiple with one map for each year in the data. The following chunk of code is similar to the single state-level maps we have drawn so far. We specify the map data as usual, adding `geom_polygon()` and `coord_map()` to it, with the arguments those functions need. Instead of cutting our data into bins, we will plot the continuous values for the adjusted death rate variable (`adjusted`) directly. To help plot this variable effectively, we will use a new scale function from the `viridis` library. The viridis colors run in low-to-high sequences and do a very good job of combining perceptually uniform colors with easy-to-see, easily contrasted hues along their scales. The `viridis` library provides continuous and discrete versions, both in several alternatives. Some balanced palettes can be a little washed out at their lower end, especially, but the viridis palettes avoid this. In this code, the `_c` suffix in the `scale_fill_viridis_c()` function signals that it is the scale for continuous data. There is a `scale_fill_viridis_d()` equivalent for discrete data.

We facet the maps just like any other small-multiple with `facet_wrap()`. We use the `theme()` function to put the legend at the bottom and remove the default shaded background from the year labels. We will learn more about this use of the `theme()` function in chapter 8. The final map is shown in figure 7.16.

If you want to experiment with cutting the data in to groups, take a look at the `cut_interval()` function.

```
library(viridis)

p0 ← ggplot(data = subset(opiates_map, year > 1999),
            mapping = aes(x = long, y = lat,
                group = group,
                fill = adjusted))

p1 ← p0 + geom_polygon(color = "gray90", size = 0.05) +
    coord_map(projection = "albers", lat0 = 39, lat1 = 45)

p2 ← p1 + scale_fill_viridis_c(option = "plasma")

p2 + theme_map() + facet_wrap(~ year, ncol = 3) +
```

Opiate related deaths by state, 2000–2014

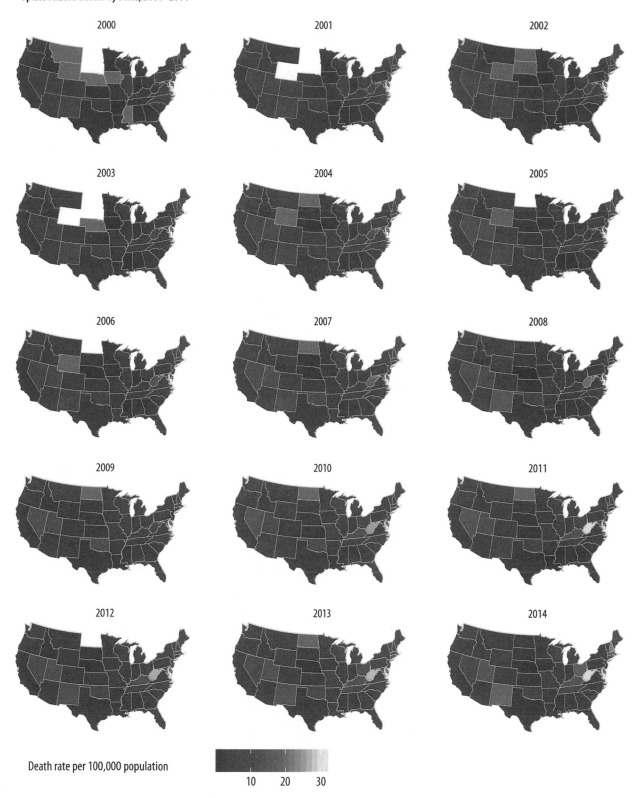

Death rate per 100,000 population

10   20   30

Figure 7.16: A small-multiple map. States in gray recorded too low a number of deaths to reliably calculate a death rate for that year. States in white reported no data.

```
    theme(legend.position = "bottom",
        strip.background = element_blank()) +
    labs(fill = "Death rate per 100,000 population ",
        title = "Opiate Related Deaths by State, 2000-2014")
```

Try revisiting your code for the ur-choropleths, but use continuous rather than binned measures, as well as the viridis palette. Instead of pct_black, use the black variable. For the population density, divide pop by land_area. You will need to adjust the scale_ functions. How do the maps compare to the binned versions? What happens to the population density map, and why?

Is this a good way to visualize this data? As we discussed above, choropleth maps of the United States tend to track first the size of the local population and secondarily the percent of the population that is African American. The differences in the geographical size of states makes spotting changes more difficult again. And it is quite difficult to compare repeatedly across spatial regions. The repeated measures do mean that some comparison is possible, and the strong trends for this data make things a little easier to see. In this case, a casual viewer might think, for example, that the opioid crisis was worst in the desert Southwest in comparison to many other parts of the country, although it also seems that something serious is happening in the Appalachians.

7.5 Is Your Data Really Spatial?

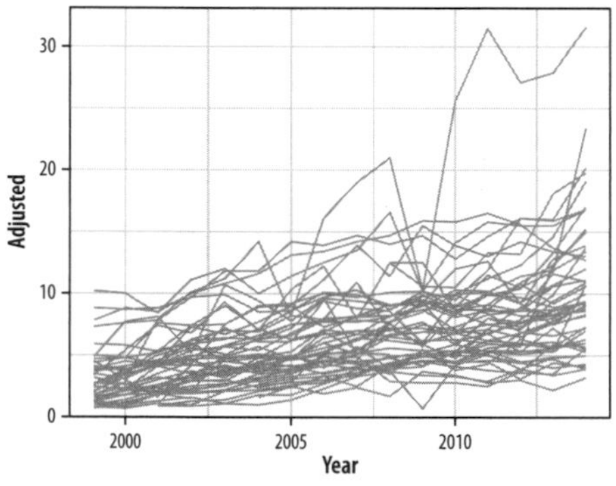

As we noted at the beginning of the chapter, even if our data is collected via or grouped into spatial units, it is always worth asking whether a map is the best way to present it. Much county, state, and national data is not properly spatial, insofar as it is really about individuals (or some other unit of interest) rather than the geographical distribution of those units per se. Let's take our state-level opiates data and redraw it as a time-series plot. We will keep the state-level focus (these are state-level rates, after all) but try to make the trends more directly visible.

We could just plot the trends for every state, as we did at the very beginning with the gapminder data. But fifty states (as in fig. 7.17) is too many lines to keep track of at once.

Figure 7.17: All the states at once.

```
p ← ggplot(data = opiates, mapping = aes(x = year, y = adjusted,
    group = state))
p + geom_line(color = "gray70")
```

A more informative approach is to take advantage of the geographical structure of the data by using the census regions to group the states. Imagine a faceted plot showing state-level trends within each region of the country, perhaps with a trend line for each region. To do this, we will take advantage of ggplot's ability to layer geoms one on top of another, using a different dataset in each case. We begin by taking the opiates data (removing Washington, DC, as it is not a state) and plotting the adjusted death rate over time.

```
p0 ← ggplot(data = drop_na(opiates, division_name),
            mapping = aes(x = year, y = adjusted))

p1 ← p0 + geom_line(color = "gray70",
             mapping = aes(group = state))
```

The drop_na() function deletes rows that have observations missing on the specified variables, in this case just division_name, because Washington, DC, is not part of any census division. We map the group aesthetic to state in geom_line(), which gives us a line plot for every state. We use the color argument to set the lines to a light gray. Next we add a smoother:

```
p2 ← p1 + geom_smooth(mapping = aes(group = division_name),
                      se = FALSE)
```

For this geom we set the group aesthetic to division_name. (Division is a smaller census classification than region.) If we set it to state, we would get fifty separate smoothers in addition to our fifty trend lines. Then, using what we learned in chapter 4, we add a geom_text_repel() object that puts the label for each state at the end of the series. Because we are labeling lines rather than points, we want the state label to appear only at the end of the line. The trick is to subset the data so that only the points from the last year observed are used (and thus labeled). We also must remember to remove Washington, DC, again here, as the new data argument supersedes the original one in p0.

```
p3 ← p2 + geom_text_repel(data = subset(opiates,
                                         year == max(year) & abbr !="DC"),
                 mapping = aes(x = year, y = adjusted, label = abbr),
                 size = 1.8, segment.color = NA, nudge_x = 30) +
      coord_cartesian(c(min(opiates$year),
            max(opiates$year)))
```

By default, geom_text_repel will add little line segments that indicate what the labels refer to. But that is not helpful here, as we are already dealing with the end point of a line. So we turn them off with the argument segment.color = NA. We also bump the labels off to the right of the lines a little, using the nudge_x argument, and use coord_cartesian() to set the axis limits so that there is enough room for them.

Finally, we facet the results by census division and add our labels. A useful adjustment is to reorder the panels by the average death rate. We put a minus in front of adjusted so that the divisions with the highest average rates appear in the chart first.

```
p3 + labs(x = "", y = "Rate per 100,000 population",
       title = "State-Level Opiate Death Rates by Census Division, 1999-2014") +
    facet_wrap(~ reorder(division_name, -adjusted, na.rm = TRUE), nrow  = 3)
```

Our new plot (fig. 7.18) brings out much of the overall story that is in the maps but also shifts the emphasis a bit. It is easier to see more clearly what is happening in some parts of the country. In particular you can see the climbing numbers in New Hampshire, Rhode Island, Massachussetts, and Connecticut. You can more easily see the state-level differences in the West, for instance between Arizona, on the one hand, and New Mexico or Utah on the other. And as was also visible on the maps, the astonishingly rapid rise in West Virginia's death rate is also evident. Finally, the time-series plots are better at conveying the diverging trajectories of various states within regions. There is more variance at the end of the series than at the beginning, especially in the Northeast, Midwest, and South, and while this can be inferred from the maps, it is easier to see in the trend plots.

The unit of observation in this graph is still the state-year. The geographically bound nature of the data never goes away. The lines we draw still represent states. Thus the basic arbitrariness of the representation cannot be made to disappear. In some sense, an ideal dataset here would be collected at some much more fine-grained level of unit, time, and spatial specificity. Imagine individual-level data with arbitrarily precise information on personal characteristics, time, and location of death. In a case like that, we could then aggregate up to any categorical, spatial, or temporal units we liked. But data like that is extremely rare, often for good reasons that range from practicality of collection to the

State-level opiate death rates by census division, 1999–2014

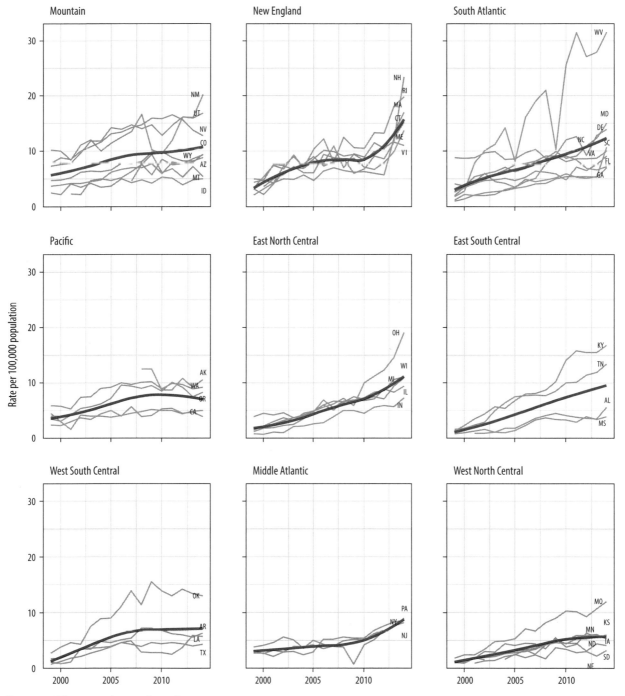

Figure 7.18: The opiate data as a faceted time-series.

privacy of individuals. In practice we need to take care not to commit a kind of fallacy of misplaced concreteness that mistakes the unit of observation for the thing of real substantive or theoretical interest. This is a problem for most kinds of social-scientific data. But their striking visual character makes maps perhaps more vulnerable to this problem than other kinds of visualization.

7.6 Where to Go Next

In this chapter, we learned how to begin to work with state-level and county-level data organized by FIPS codes. But this barely scratches the surface of visualization where spatial features and distributions are the main focus. The analysis and visualization of spatial data is its own research area, with its own research disciplines in geography and cartography. Concepts and methods for representing spatial features are both well developed and standardized. Until recently, most of this functionality was accessible only through dedicated Geographical Information Systems. Their mapping and spatial analysis features were not well connected. Or, at least, they were not conveniently connected to software oriented to the analysis of tabular data.

This is changing fast. Brundson & Comber (2015) provide an introduction to some of R's mapping capabilities. Meanwhile, recently these tools have become much more accessible via the tidyverse. Of particular interest to social scientists is Edzer Pebesma's ongoing development of the sf package, which implements the standard Simple Features data model for spatial features in a tidyverse-friendly way. Relatedly, Kyle Walker and Bob Rudis's tigris package allows for (sf-library-compatible) access to the U.S. Census Bureau's TIGER/Line shapefiles, which allow you to map data for many different geographical, administrative, and census-related subdivisions of the United States, as well as things like roads and water features. Finally, Kyle Walker's tidycensus package (Walker 2018) makes it much easier to tidily get both substantive and spatial feature data from the U.S. Census and the American Community Survey.

r-spatial.github.io/sf/. Also see news and updates at r-spatial.org.

github.com/walkerke/tigris

walkerke.github.io/tidycensus

8 Refine Your Plots

So far we have mostly used ggplot's default output when making our plots, generally not looking at opportunities to tweak or customize things to any great extent. In general, when making figures during exploratory data analysis, the default settings in ggplot should be pretty good to work with. It's only when we have some specific plot in mind that the question of polishing the results comes up. Refining a plot can mean several things. We might want to get the look of it just right, based on our own tastes and our sense of what needs to be highlighted. We might want to format it in a way that will meet the expectations of a journal, a conference audience, or the general public. We might want to tweak this or that feature of the plot or add an annotation or additional detail not covered by the default output. Or we might want to completely change the look of the entire thing, given that all the structural elements of the plot are in place. We have the resources in ggplot to do all these things.

Let's begin by looking at a new dataset, asasec. This is some data on membership over time in special-interest sections of the American Sociological Association.

```
head(asasec)
```

```
##                                 Section         Sname
## 1       Aging and the Life Course (018)         Aging
## 2     Alcohol, Drugs and Tobacco (030) Alcohol/Drugs
## 3  Altruism and Social Solidarity (047)      Altruism
## 4              Animals and Society (042)       Animals
## 5              Asia/Asian America (024)          Asia
## 6              Body and Embodiment (048)          Body
##   Beginning Revenues Expenses Ending Journal Year Members
## 1     12752    12104    12007  12849      No 2005     598
## 2     11933     1144      400  12677      No 2005     301
## 3      1139     1862     1875   1126      No 2005      NA
## 4       473      820     1116    177      No 2005     209
```

```
## 5      9056    2116    1710    9462    No 2005    365
## 6      3408    1618    1920    3106    No 2005    NA
```

In this dataset, we have membership data for each section over a ten-year period, but the data on section reserves and income (the Beginning and Revenues variables) is for 2015 only. Let's look at the relationship between section membership and section revenues for a single year, 2014.

```
p ← ggplot(data = subset(asasec, Year == 2014), mapping = aes(x = Members,
    y = Revenues, label = Sname))

p + geom_point() + geom_smooth()
```

```
## `geom_smooth()` using method = 'loess' and formula 'y ~ x'
```

Figure 8.1 is our basic scatterplot-and-smoother graph. To refine it, let's begin by identifying some outliers, switch from loess to OLS, and introduce a third variable. This gets us to Figure 8.2.

```
p ← ggplot(data = subset(asasec, Year == 2014), mapping = aes(x = Members,
    y = Revenues, label = Sname))

p + geom_point(mapping = aes(color = Journal)) + geom_smooth(method = "lm")
```

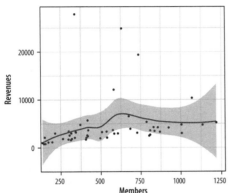

Figure 8.1: Back to basics.

Now we can add some text labels. At this point it makes sense to use some intermediate objects to build things up as we go. We won't show them all. But by now you should be able to see in your mind's eye what an object like p1 or p2 will look like. And of course you should type out the code and check if you are right as you go.

```
p0 ← ggplot(data = subset(asasec, Year == 2014), mapping = aes(x = Members,
    y = Revenues, label = Sname))

p1 ← p0 + geom_smooth(method = "lm", se = FALSE, color = "gray80") +
    geom_point(mapping = aes(color = Journal))

p2 ← p1 + geom_text_repel(data = subset(asasec, Year == 2014 &
    Revenues > 7000), size = 2)
```

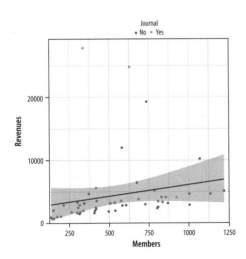

Figure 8.2: Refining the plot.

Continuing with the p2 object, we can label the axes and scales. We also add a title and move the legend to make better use of the space in the plot (fig. 8.3).

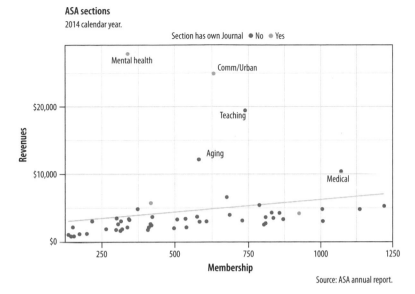

```
p3 ← p2 + labs(x="Membership",
        y="Revenues",
        color = "Section has own Journal",
        title = "ASA Sections",
        subtitle = "2014 Calendar year.",
        caption = "Source: ASA annual report.")
p4 ← p3 + scale_y_continuous(labels = scales::dollar) +
    theme(legend.position = "bottom")
p4
```

8.1 Use Color to Your Advantage

You should choose a color palette in the first place based on its ability to express the data you are plotting. An unordered categorical variable like "country" or "sex," for example, requires distinct colors that won't be easily confused with one another. An ordered categorical variable like "level of education," on the other hand, requires a graded color scheme of some kind running from less to more or earlier to later. There are other considerations, too. For example, if your variable is ordered, is your scale centered on a neutral midpoint with departures to extremes in each direction, as in a Likert scale? Again, these questions are about

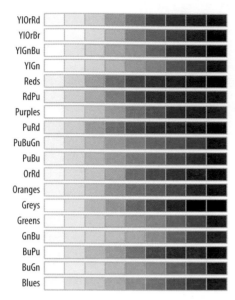

Figure 8.4: RColorBrewer's sequential palettes.

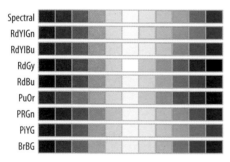

Figure 8.5: RColorBrewer's diverging palettes.

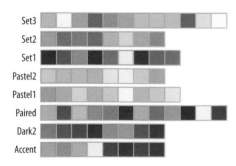

Figure 8.6: RColorBrewer's qualitative palettes.

ensuring accuracy and fidelity when mapping a variable to a color scale. Take care to choose a palette that reflects the structure of your data. For example, do not map sequential scales to categorical palettes, or use a diverging palette for a variable with no well-defined midpoint.

Separate from these mapping issues, there are considerations about which colors in particular to choose. In general, the default color palettes that ggplot makes available are well chosen for their perceptual properties and aesthetic qualities. We can also use color and color layers as device for emphasis, to highlight particular data points or parts of the plot, perhaps in conjunction with other features.

We choose color palettes for mappings through one of the scale_ functions for color or fill. While it is possible to very finely control the look of your color schemes by varying the hue, chroma, and luminance of each color you use via scale_color_hue() or scale_fill_hue(), in general this is not recommended. Instead you should use the RColorBrewer package to make a wide range of named color palettes available to you, and choose from those. Figures 8.4, 8.5, and 8.6 show the available options for sequential, diverging, and qualitative variables. When used in conjunction with ggplot, you access these colors by specifying the scale_color_brewer() or scale_fill_brewer() functions, depending on the aesthetic you are mapping. Figure 8.7 shows how to use the named palettes in this way.

```
p ← ggplot(data = organdata, mapping = aes(x = roads, y = donors,
    color = world))
p + geom_point(size = 2) + scale_color_brewer(palette = "Set2") +
    theme(legend.position = "top")


p + geom_point(size = 2) + scale_color_brewer(palette = "Pastel2") +
    theme(legend.position = "top")


p + geom_point(size = 2) + scale_color_brewer(palette = "Dark2") +
    theme(legend.position = "top")
```

You can also specify colors manually, via scale_color_manual() or scale_fill_manual(). These functions take a value argument that can be specified as vector of color names or color values that R knows about. R knows many color names (like `red`, and `green`, and `cornflowerblue`). Try `demo('colors')` for an overview. Alternatively, color values can be specified via their hexadecimal RGB value. This is a way of encoding color values in the RGB colorspace, where each channel can take a value from 0 to 255. A color hex value begins with a hash or pound character, #, followed by three pairs of hexadecimal or "hex" numbers. Hex values are in Base 16, with the first six letters of the alphabet standing for the numbers 10 to 15. This allows a two-character hex number to range from 0 to 255. You read them as `#rrggbb`, where `rr` is the two-digit hex code for the red channel, `gg` for the green channel, and `bb` for the blue channel. So `#CC55DD` translates in decimal to `CC` = 204 (red), `55` = 85 (green), and `DD` = 221 (blue). It gives a strong pink color.

Going back to our ASA membership plot, for example, for figure 8.8 we can manually introduce a palette from Chang (2013) that's friendly to viewers who are color blind.

```
cb_palette ← c("#999999", "#E69F00", "#56B4E9", "#009E73", "#F0E442",
    "#0072B2", "#D55E00", "#CC79A7")

p4 + scale_color_manual(values = cb_palette)
```

While we can specify colors manually, this work has already been done for us. If we are serious about using a safe palette for color-blind viewers, we should investigate the `dichromat` package instead. It provides a range of palettes and some useful functions for helping you see approximately what your current palette might look like to a viewer with one of several different kinds of color blindness.

For example, let's use RColorBrewer's `brewer.pal()` function to get five colors from ggplot's default palette.

```
Default ← brewer.pal(5, "Set2")
```

Next, we can use a function from the `dichromat` package to transform these colors to new values that simulate different kinds of color blindness. The results are shown in figure 8.9.

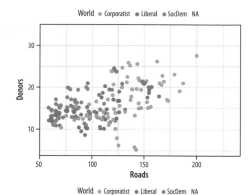

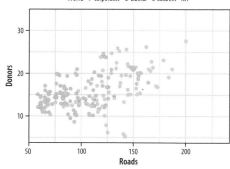

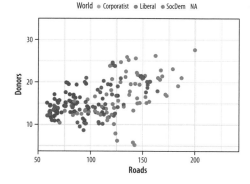

Figure 8.7: Some available palettes in use.

The `colorblindr` package has similar functionality.

Figure 8.8: Using a custom color palette.

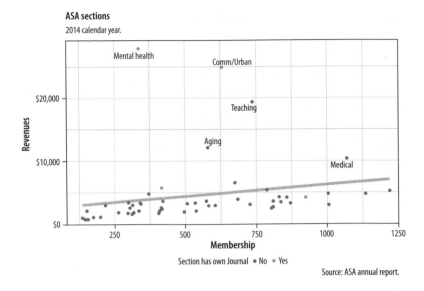

```r
library(dichromat)


types ← c("deutan", "protan", "tritan")
names(types) ← c("Deuteronopia", "Protanopia", "Tritanopia")


color_table ← types %>% purrr::map(~dichromat(Default, .x)) %>%
    as_tibble() %>% add_column(Default, .before = TRUE)


color_table
```

```
## # A tibble: 5 x 4
##   Default Deuteronopia Protanopia Tritanopia
##   <chr>   <chr>        <chr>      <chr>
## 1 #66C2A5 #AEAEA7      #BABAA5    #82BDBD
## 2 #FC8D62 #B6B661      #9E9E63    #F29494
## 3 #8DA0CB #9C9CCB      #9E9ECB    #92ABAB
## 4 #E78AC3 #ACACC1      #9898C3    #DA9C9C
## 5 #A6D854 #CACA5E      #D3D355    #B6C8C8
```

```r
color_comp(color_table)
```

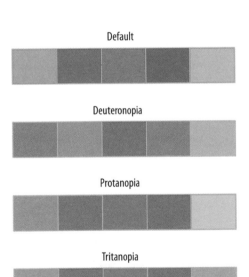

Figure 8.9: Comparing a default color palette with an approximation of how the same palette appears to people with one of three kinds of color blindness.

In this code, we create a vector of types of color blindness that the dichromat() function knows about and give them proper names. Then we make a table of colors for each type using the purrr library's map() function. The rest of the pipeline converts the results from a list to a tibble and adds the original colors as the first column in the table. We can now plot them to see how they compare, using a convenience function from the socviz library.

The ability to manually specify colors can be useful when the meaning of a category itself has a strong color association. Political parties, for example, tend to have official or quasi-official party colors that people associate with them. In such cases it is helpful to be able to present results for, say, the Green Party in a green color. When doing this, it is worth keeping in mind that some colors are associated with categories (especially categories of person) for outmoded reasons, or no very good reason. Do not use stereotypical colors just because you can.

8.2 Layer Color and Text Together

Aside from mapping variables directly, color is also useful when we want to pick out or highlight some aspect of our data. In cases like this, the layered approach of ggplot can really work to our advantage. Let's work through an example where we use manually specified colors both for emphasis and because of their social meaning.

We will build up a plot of data about the U.S. general election in 2016. It is contained in the county_data object in the socviz library. We begin by defining a blue and red color for the Democrats and Republicans, respectively. Then we create the basic setup and first layer of the plot. We subset the data, including only counties with a value of "No" on the flipped variable. We set the color of geom_point() to be a light gray, as it will form the background layer of the plot (fig. 8.10). And we apply a log transformation to the x-axis scale.

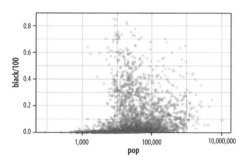

Figure 8.10: The background layer.

```
# Democrat Blue and Republican Red
party_colors ← c("#2E74C0", "#CB454A")


p0 ← ggplot(data = subset(county_data,
                          flipped == "No"),
            mapping = aes(x = pop,
                          y = black/100))


p1 ← p0 + geom_point(alpha = 0.15, color = "gray50") +
    scale_x_log10(labels=scales::comma)


p1
```

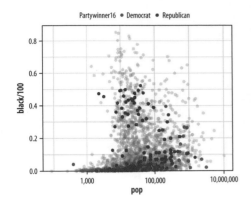

Figure 8.11: The second layer.

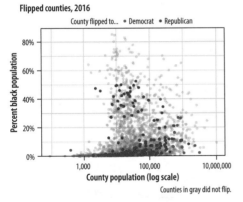

Figure 8.12: Adding guides and labels, and fixing the y-axis scale.

In the next step (fig. 8.11) we add a second `geom_point()` layer. Here we start with the same dataset but extract a complementary subset from it. This time we choose the "Yes" counties on the `flipped` variable. The `x` and `y` mappings are the same, but we add a color scale for these points, mapping the `partywinner16` variable to the `color` aesthetic. Then we specify a manual color scale with `scale_color_manual()`, where the values are the blue and red `party_colors` we defined above.

```
p2 ← p1 + geom_point(data = subset(county_data,
                                   flipped == "Yes"),
                     mapping = aes(x = pop, y = black/100,
                                   color = partywinner16)) +
    scale_color_manual(values = party_colors)

p2
```

The next layer sets the y-axis scale and the labels (fig. 8.12).

```
p3 ← p2 + scale_y_continuous(labels=scales::percent) +
    labs(color = "County flipped to ... ",
         x = "County Population (log scale)",
         y = "Percent Black Population",
         title = "Flipped counties, 2016",
         caption = "Counties in gray did not flip.")

p3
```

Finally, we add a third layer using the `geom_text_repel()` function. Once again we supply a set of instructions to subset the data for this text layer. We are interested in the flipped counties that have a relatively high percentage of African American residents. The result, shown in figure 8.13, is a complex but legible multilayer plot with judicious use of color for variable coding and context.

```
p4 ← p3 + geom_text_repel(data = subset(county_data,
                                        flipped == "Yes" &
                                        black > 25),
```

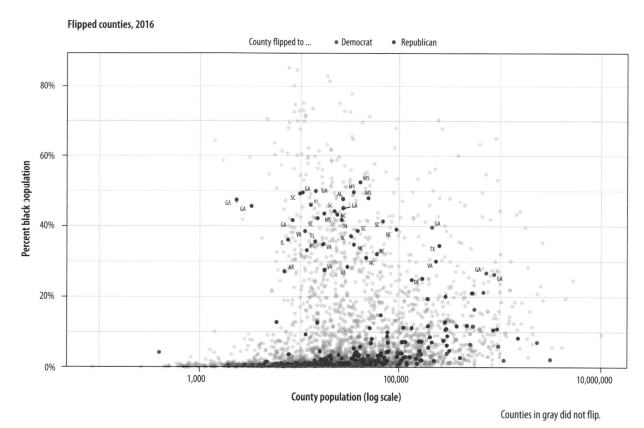

Figure 8.13: County-level election data from 2016.

```
              mapping = aes(x = pop,
                            y = black/100,
                            label = state), size = 2)

p4 + theme_minimal() +
    theme(legend.position="top")
```

When producing a graphic like this in ggplot, or when looking at good plots made by others, it should gradually become your
habit to see not just the content of the plot but also the implicit
or explicit structure that it has. First, you will be able to see the
mappings that form the basis of the plot, picking out which variables are mapped to x and y, and which to to color, fill, shape,
label, and so on. What geoms were used to produce them? Second, how have the scales been adjusted? Are the axes transformed?

Are the fill and color legends combined? And third, especially as you practice making plots of your own, you will find yourself picking out the *layered* structure of the plot. What is the base layer? What has been drawn on top of it, and in what order? Which upper layers are formed from subsets of the data? Which are new datasets? Are there annotations? The ability to evaluate plots in this way, to apply the grammar of graphics in practice, is useful both for looking at plots and for thinking about how to make them.

8.3 Change the Appearance of Plots with Themes

Our elections plot is in a pretty finished state. But if we want to change the overall look of it all at once, we can do that using ggplot's theme engine. Themes can be turned on or off using the theme_set() function. It takes the name of a theme (which will itself be a function) as an argument. Try the following:

```
theme_set(theme_bw())
p4 + theme(legend.position = "top")

theme_set(theme_dark())
p4 + theme(legend.position = "top")
```

Internally, theme functions are a set of detailed instructions to turn on, turn off, or modify a large number of graphical elements on the plot. Once set, a theme applies to all subsequent plots, and it remains active until it is replaced by a different theme. This can be done either through the use of another theme_set() statement or on a per plot basis by adding the theme function to the end of the plot: p4 + theme_gray() would temporarily override the generally active theme for the p4 object only. You can still use the theme() function to fine-tune any aspect of your plot, as seen above with the relocation of the legend to the top of the graph.

The ggplot library comes with several built-in themes, including theme_minimal() and theme_classic(), with theme_gray() or theme_grey() as the default. If these are not to your taste, install the ggthemes package for many more options. You can, for example, make ggplot output look like it has been featured in the

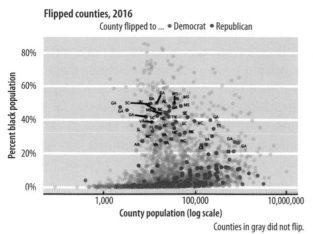
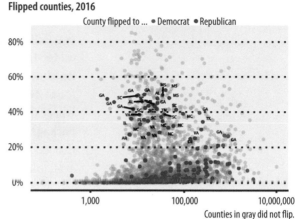

Figure 8.14: *Economist* and *WSJ* themes.

Economist, or the *Wall Street Journal*, or in the pages of a book by Edward Tufte. Figure 8.14 shows two examples.

Using some themes might involve adjusting font sizes or other elements as needed, if the defaults are too large or small. If you use a theme with a colored background, you will also need to consider what color palette you are using when mapping to color or fill aesthetics. You can define your own themes either entirely from scratch or by starting with one you like and making adjustments from there.

```
library(ggthemes)

theme_set(theme_economist())
p4 + theme(legend.position="top")

theme_set(theme_wsj())

p4 + theme(plot.title = element_text(size = rel(0.6)),
           legend.title = element_text(size = rel(0.35)),
           plot.caption = element_text(size = rel(0.35)),
           legend.position = "top")
```

Generally speaking, themes with colored backgrounds and customized typefaces are best used when making one-off graphics or posters, preparing figures to integrate into a slide presentation, or conforming to a house or editorial style for publication. Take

It also contains some convenience functions for laying out several plot objects in a single figure, among other features, as we shall see below in one of the case studies.

care to consider how the choices you make will harmonize with the broader printed or displayed material. Just as with the choice of palettes for aesthetic mappings, when starting out it can be wisest to stick to the defaults or consistently use a theme that has had its kinks already ironed out. Claus O. Wilke's cowplot package, for instance, contains a well-developed theme suitable for figures whose final destination is a journal article. Bob Rudis's hrbrthemes package, meanwhile, has a distinctive and compact look and feel that takes advantage of some freely available typefaces. Both are available via install.packages().

The theme() function allows you to exert fine-grained control over the appearance of all kinds of text and graphical elements in a plot. For example, we can change the color, typeface, and font size of text. If you have been following along writing your code, you will have noticed that the plots you make have not been identical to the ones shown in the text. The axis labels are in a slightly different place from the default, the typeface is different, and there are other, smaller changes as well. The theme_book() function provides the custom ggplot theme used throughout this book. The code for this theme is based substantially on Bob Rudis's theme_ipsum(), from his hrbrthemes library. You can learn more about it in the appendix. For figure 8.15, we then adjust that theme even further by tweaking the text size, and we also remove a number of elements by naming them and making them disappear using element_blank().

```
p4 + theme(legend.position = "top")


p4 + theme(legend.position = "top",
           plot.title = element_text(size=rel(2),
                                     lineheight=.5,
                                     family="Times",
                                     face="bold.italic",
                                     colour="orange"),
           axis.text.x = element_text(size=rel(1.1),
                                      family="Courier",
                                      face="bold",
                                      color="purple"))
```

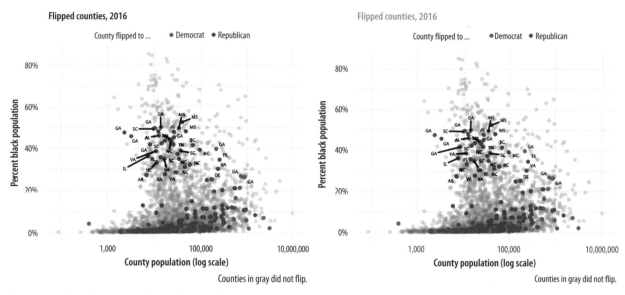

Figure 8.15: Controlling various theme elements directly (and making several bad choices while doing so).

8.4 Use Theme Elements in a Substantive Way

It makes good sense to use themes as a way to fix design elements because that means you can subsequently ignore them and focus instead on the data you are examining. But it is also worth remembering that ggplot's theme system is very flexible. It permits a wide range of design elements to be adjusted to create custom figures. For instance, following an example from Wehrwein (2017), we will create an effective small multiple of the age distribution of GSS respondents over the years. The gss_lon data contains information on the age of each GSS respondent for all the years in the survey since 1972. The base figure 8.16 is a scaled geom_density() layer of the sort we saw earlier, this time faceted by the year variable. We will fill the density curves with a dark gray color and then add an indicator of the mean age in each year, and a text layer for the label. With those in place we then adjust the detail of several theme elements, mostly to remove them. As before, we use element_text() to tweak the appearance of various text elements such as titles and labels. And we also use element_blank() to remove several of them altogether.

First, we need to calculate the mean age of the respondents for each year of interest. Because the GSS has been around for most (but not all) years since 1972, we will look at distributions about

Age distribution of GSS respondents

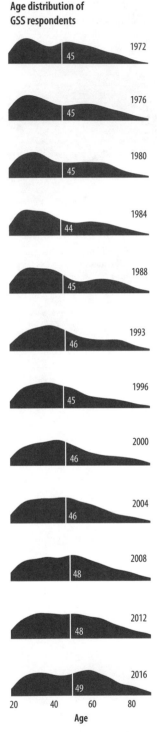

Figure 8.16: A customized small multiple.

every four years since the beginning. We use a short pipeline to extract the average ages.

```
yrs ← c(seq(1972, 1988, 4), 1993, seq(1996, 2016, 4))

mean_age ← gss_lon %>%
    filter(age %nin% NA && year %in% yrs) %>%
    group_by(year) %>%
    summarize(xbar = round(mean(age, na.rm = TRUE), 0))
mean_age$y ← 0.3

yr_labs ← data.frame(x = 85, y = 0.8,
                        year = yrs)
```

The y column in mean_age will come in handy when we want to position the age as a text label. Next, we prepare the data and set up the geoms.

```
p ← ggplot(data = subset(gss_lon, year %in% yrs),
          mapping = aes(x = age))

p1 ← p + geom_density(fill = "gray20", color = FALSE,
                        alpha = 0.9, mapping = aes(y = ..scaled..)) +
    geom_vline(data = subset(mean_age, year %in% yrs),
              aes(xintercept = xbar), color = "white", size = 0.5) +
    geom_text(data = subset(mean_age, year %in% yrs),
             aes(x = xbar, y = y, label = xbar), nudge_x = 7.5,
             color = "white", size = 3.5, hjust = 1) +
    geom_text(data = subset(yr_labs, year %in% yrs),
             aes(x = x, y = y, label = year)) +
    facet_grid(year ~ ., switch = "y")
```

The initial p object subsets the data by the years we have chosen and maps x to the age variable. The geom_density() call is the base layer, with arguments to turn off its default line color, set the fill to a shade of gray, and scale the y-axis between zero and one.

Using our summarized dataset, the geom_vline() layer draws a vertical white line at the mean age of the distribution. The first of two text geoms labels the age line (in white). The first geom_text()

call uses a nudge argument to push the label slightly to the right of its x-value. The second labels the year. We do this because we are about to turn off the usual facet labels to make the plot more compact. Finally we use facet_grid() to break out the age distributions by year. We use the switch argument to move the labels to the left.

With the structure of the plot in place, we then style the elements in the way that we want, using a series of instructions to theme().

```
p1 + theme_book(base_size = 10, plot_title_size = 10,
                strip_text_size = 32, panel_spacing = unit(0.1, "lines")) +
    theme(plot.title = element_text(size = 16),
          axis.text.x= element_text(size = 12),
          axis.title.y=element_blank(),
          axis.text.y=element_blank(),
          axis.ticks.y = element_blank(),
          strip.background = element_blank(),
          strip.text.y = element_blank(),
          panel.grid.major = element_blank(),
          panel.grid.minor = element_blank()) +
    labs(x = "Age",
         y = NULL,
         title = "Age Distribution of\nGSS Respondents")
```

One of the pleasing things about ggplot's developer community is that it often takes plot ideas that are first worked out in a one-off or bespoke way and generalizes them to the point where they are available as new geoms. Shortly after writing the code for the GSS age distributions in figure 8.16, the ggridges package was released. Written by Claus O. Wilke, it offers a different take on small-multiple density plots by allowing the distributions to overlap vertically to interesting effect. It is especially useful for repeated distributional measures that change in a clear direction. In figure 8.17 we redo our previous plot using a function from ggridges. Because geom_density_ridges() makes for a more compact display we trade off showing the mean age value for the sake of displaying the distribution for every GSS year.

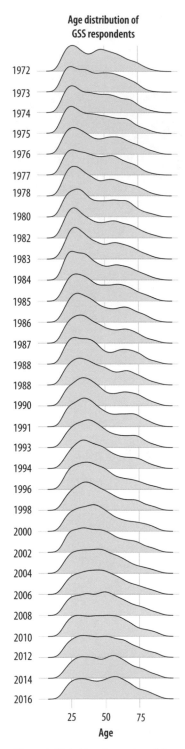

**Age distribution of
GSS respondents**

1972
1973
1974
1975
1976
1977
1978
1980
1982
1983
1984
1985
1986
1987
1988
1988
1990
1991
1993
1994
1996
1998
2000
2002
2004
2006
2008
2010
2012
2014
2016

25   50   75
Age

Figure 8.17: A ridgeplot version of the age distribution plot.

```
library(ggridges)

p ← ggplot(data = gss_lon,
           mapping = aes(x = age, y = factor(year, levels = rev(unique(year)),
                                             ordered = TRUE)))

p + geom_density_ridges(alpha = 0.6, fill = "lightblue", scale = 1.5) +
    scale_x_continuous(breaks = c(25, 50, 75)) +
    scale_y_discrete(expand = c(0.01, 0)) +
    labs(x = "Age", y = NULL,
         title = "Age Distribution of\nGSS Respondents") +
    theme_ridges() +
    theme(title = element_text(size = 16, face = "bold"))
```

The expand argument in scale_y_discrete() adjusts the scaling of the y-axis slightly. It has the effect of shortening the distance between the axis labels and the first distribution, and it also prevents the top of the very first distribution from being cut off by the frame of the plot. The package also comes with its own theme, theme_ridges(), that adjusts the labels so that they are aligned properly, and we use it here. The geom_density_ridges() function is also capable of reproducing the look of our original version. The degree of overlap in the distributions is controlled by the scale argument in the geom. You can experiment with setting it to values below or above one to see the effects on the layout of the plot.

Much more detailed information on the names of the various elements you can control via theme() can be found in the ggplot documentation. Setting these thematic elements in an ad hoc way is often one of the first things people want to do when they make a plot. But in practice, apart from getting the overall size and scale of your plot squared away, making small adjustments to theme elements should be the last thing you do in the plotting process. Ideally, once you have set up a theme that works well for you, it should be something you can avoid having to do at all.

8.5 Case Studies

Bad graphics are everywhere. Better ones are within our reach. For the final sections of this chapter we will work through some common visualization problems or dilemmas, as seen through some real-life cases. In each case we will look at the original figures and redraw them in new (and better) versions. In the process we will introduce a few new functions and features of ggplot that we have not seen yet. This, too, is true to life. Usually, it's having to face some practical design or visualization question that forces us to ferret out the solution to our problem in the documentation or come up with some alternative answer on the fly ourselves. Let's start with a common case: the use of dual axes in trend plots.

Two y-axes

In January 2016 Liz Ann Sonders, chief investment strategist with Charles Schwab, Inc., tweeted about the apparent correlation between two economic time series: the Standard and Poor's 500 stock market index and the Monetary Base, a measure of the size of money supply. The S&P is an index that ranges from about 700 to about 2,100 over the period of interest (about the last seven years). The Monetary Base ranges from about 1.5 trillion to 4.1 trillion dollars over the same period. This means that we can't plot the two series directly. The Monetary Base is so much larger that it would make the S&P 500 series appear as a flat line at the bottom. While there are several reasonable ways to address this, people often opt instead to have two y-axes.

Because it is designed by responsible people, R makes it slightly tricky to draw graphs with two y-axes. In fact, ggplot rules it out of order altogether. It is possible to do it using R's base graphics, if you insist. Figure 8.18 shows the result. (You can find the code at https://github.com/kjhealy/two-y-axes) Graphics in base R work very differently from the approach we have taken throughout this book, so it would just be confusing to show the code here.

Most of the time when people draw plots with two y-axes they want to line the series up as closely as possible because they suspect that there's a substantive association between them, as in this

Figure 8.18: Two time series, each with its own y-axis.

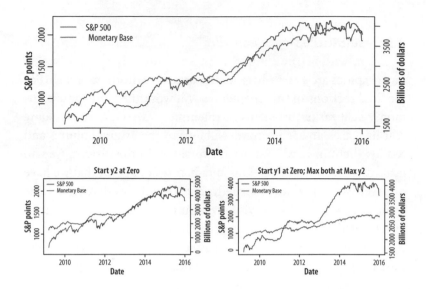

Figure 8.19: Variations on two y-axes.

case. The main problem with using two y-axes is that it makes it even easier than usual to fool yourself (or someone else) about the degree of association between the variables. This is because you can adjust the scaling of the axes relative to one another in a way that moves the data series around more or less however you like. In figure 8.18 the red Monetary Base line tracks below the blue S&P 500 for the first half of the graph and is above it for the second half.

We can "fix" that by deciding to start the second y-axis at zero, which shifts the Monetary Base line above the S&P line for the first half of the series and below it later on. The first panel in figure 8.19 shows the results. The second panel, meanwhile, adjusts the axes so that the axis tracking the S&P starts at zero. The axis tracking the Monetary Base starts around its minimum (as is generally good practice), but now both axes max out around 4,000. The units are different, of course. The 4,000 on the S&P side is an index number, while the Monetary Base number is 4,000 billion dollars. The effect is to flatten out the S&P's apparent growth quite a bit, muting the association between the two variables substantially. You could tell quite a different story with this one, if you felt like it.

How else might we draw this data? We could use a split- or broken-axis plot to show the two series at the same time. These can be effective sometimes, and they seem to have better perceptual properties than overlayed charts with dual axes (Isenberg et al. 2011). They are most useful in cases where the series you are plotting are of the same kind but of very different magnitudes. That is not the case here.

Another compromise, if the series are not in the same units (or of widely differing magnitudes), is to rescale one of the series (e.g., by dividing or multiplying it by a thousand), or alternatively to index each of them to 100 at the start of the first period and then plot them both. Index numbers can have complications of their own, but here they allow us to use one axis instead of two, and also to calculate a sensible difference between the two series and plot that as well, in a panel below. It can be tricky to visually estimate the difference between series, in part because our perceptual tendency is to look for the *nearest* comparison point in the other series rather than the one directly above or below. Following Cleveland (1994), we can also add a panel underneath that tracks the running difference between the two series. We begin by making each plot and storing them in an object. To do this, it will be convenient to tidy the data into a long format, with the indexed series in the key variable and their corresponding scores as the values. We use tidyr's gather() function for this:

```
head(fredts)
```

```
##          date   sp500  monbase sp500_i monbase_i
## 1 2009-03-11  696.68  1542228 100.000   100.000
## 2 2009-03-18  766.73  1693133 110.055   109.785
## 3 2009-03-25  799.10  1693133 114.701   109.785
## 4 2009-04-01  809.06  1733017 116.131   112.371
## 5 2009-04-08  830.61  1733017 119.224   112.371
## 6 2009-04-15  852.21  1789878 122.324   116.058
```

```
fredts_m ← fredts %>% select(date, sp500_i, monbase_i) %>% gather(key = series,
    value = score, sp500_i:monbase_i)
```

```
head(fredts_m)
```

```
##          date  series   score
## 1 2009-03-11 sp500_i 100.000
## 2 2009-03-18 sp500_i 110.055
## 3 2009-03-25 sp500_i 114.701
## 4 2009-04-01 sp500_i 116.131
## 5 2009-04-08 sp500_i 119.224
## 6 2009-04-15 sp500_i 122.324
```

Once the data is tidied in this way we can make our graph.

```
p ← ggplot(data = fredts_m,
           mapping = aes(x = date, y = score,
                              group = series,
                              color = series))
p1 ← p + geom_line() + theme(legend.position = "top") +
    labs(x = "Date",
         y = "Index",
         color = "Series")

p ← ggplot(data = fredts,
           mapping = aes(x = date, y = sp500_i - monbase_i))


p2 ← p + geom_line() +
    labs(x = "Date",
         y = "Difference")
```

Now that we have our two plots, we want to lay them out nicely. We do not want them to appear in the same plot area, but we do want to compare them. It would be possible to do this with a facet, but that would mean doing a fair amount of data munging to get all three series (the two indices and the difference between them) into the same tidy data frame. An alternative is to make two separate plots and then arrange them just as we like. For instance, have the comparison of the two series take up most of the space, and put the plot of the index differences along the bottom in a smaller area.

The layout engine used by R and ggplot, called grid, does make this possible. It controls the layout and positioning of plot areas and objects at a lower level than ggplot. Programming grid layouts directly takes a little more work than using ggplot's functions alone. Fortunately, there are some helper libraries that we can use to make things easier. One possibility is to use the gridExtra library. It provides a number of useful functions that let us talk to the grid engine, including grid.arrange(). This function takes a list of plot objects and instructions for how we would like them arranged. The cowplot library we mentioned earlier makes things even easier. It has a plot_grid() function that works much like grid.arrange() while also taking care of some fine details, including the proper alignment of axes across separate plot objects.

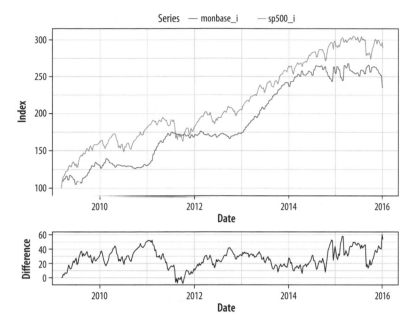

```
cowplot::plot_grid(p1, p2, nrow = 2, rel_heights = c(0.75, 0.25),
    align = "v")
```

The result is shown in figure 8.20. It looks pretty good. In this version, the S&P index runs above the Monetary Base for almost the whole series, whereas in the plot as originally drawn, they crossed.

The broader problem with dual-axis plots of this sort is that the apparent association between these variables is probably spurious. The original plot is enabling our desire to spot patterns, but it is probably the case that both of these time series are tending to increase but are not otherwise related in any deep way. If we were interested in establishing the true association between them, we might begin by naively regressing one on the other. We can try to predict the S&P index from the Monetary Base, for instance. If we do that, things look absolutely fantastic to begin with, as we appear to explain about 95 percent of the variance in the S&P just by knowing the size of the Monetary Base from the same period. We're going to be rich!

Sadly, we're probably not going to be rich. While everyone knows that correlation is not causation, with time-series data we get this problem twice over. Even just considering a single series, each observation is often pretty closely correlated with the observation in the period immediately before it, or perhaps with the observation some regular number of periods before it. A time series might have a seasonal component that we would want to account for before making claims about its growth, for example. And if we ask what *predicts* its growth, then we will introduce some other time series, which will have trend properties of its own. In those circumstances, we more or less automatically violate the assumptions of ordinary regression analysis in a way that produces wildly overconfident estimates of association. The result, which may seem paradoxical when you first run across it, is that a lot of the machinery of time-series analysis is about making the serial nature of the data go away.

Like any rule of thumb, it is possible to come up with exceptions, or talk oneself into them. We can imagine situations where the judicious use of dual y-axes might be a sensible way to present data to others, or might help a researcher productively explore a dataset. But in general I recommend against it because it is already much too easy to present spurious, or at least overconfident, associations, especially with time-series data. Scatterplots can do that just fine. Even with a single series, as we saw in chapter 1, we can make associations look steeper or flatter by fiddling with the aspect ratio. Using two y-axes gives you an extra degree of freedom to mess about with the data that, in most cases, you really should not take advantage of. A rule like this will not stop people who want to fool you with charts from trying, of course. But it might help you not fool yourself.

Redrawing a bad slide

In late 2015 Marissa Mayer's performance as CEO of Yahoo was being criticized by many observers. One of them, Eric Jackson, an investment fund manager, sent a ninety-nine-slide presentation to Yahoo's board outlining his best case against Mayer. (He also circulated it publicly.) The style of the slides was typical of business presentations. Slides and posters are a very useful means of communication. In my experience, most people who complain about "death by PowerPoint" have not sat through enough talks where the presenter hasn't even bothered to prepare slides. But it is striking to see how fully the "slide deck" has escaped

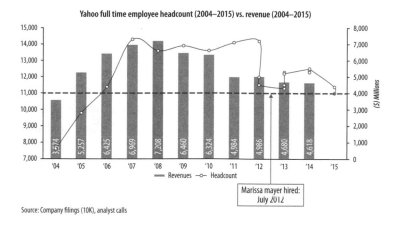

Figure 8.21: A bad slide. (From the December 2015 Yahoo! Investor Presentation: A Better Plan for Yahoo Shareholders.)

its origins as an aid to communication and metastasized into a freestanding quasi-format of its own. Business, the military, and academia have all been infected by this tendency in various ways. Never mind taking the time to write a memo or an article, just give us endless pages of bullet points and charts. The disorienting effect is of constant summaries of discussions that never took place.

In any case, figure 8.21 reproduces a typical slide from the deck. It seems to want to say something about the relationship between Yahoo's number of employees and its revenue, in the context of Mayer's tenure as CEO. The natural thing to do would be to make some kind of scatterplot to see if there was a relationship between these variables. Instead, however, the slide puts time on the x-axis and uses two y-axes to show the employee and revenue data. It plots the revenues as a bar chart and the employee data as points connected by slightly wavy lines. At first glance, it is not clear whether the connecting line segments are just manually added or if there's some principle underlying the wiggles. (They turn out to have been created in Excel.) The revenue values are used as labels within the bars. The points are not labeled. Employee data goes to 2015, but revenue data only to 2014. An arrow points to the date Mayer was hired as CEO, and a red dotted line seems to indicate … actually I'm not sure. Maybe some sort of threshold below which employee numbers should fall? Or maybe just the last observed value, to allow comparison across the series? It isn't clear. Finally, notice that while the revenue numbers are annual, there is more than one observation per year for some of the later employee numbers.

How should we redraw this chart? Let's focus on getting across the relationship between employee numbers and revenue, as that seems to be the motivation for it in the first place. As a secondary element, we want to say something about Mayer's role in this relationship. The original sin of the slide is that it plots two series of numbers using two different y-axes, as discussed above. We see this from business analysts more often than not. Time is almost the only thing they ever put on the x-axis.

To redraw the chart, I took the numbers from the bars on the chart together with employee data from QZ.com. Where there was quarterly data in the slide, I used the end-of-year number for employees, except for 2012. Mayer was appointed in July 2012. Ideally we would have quarterly revenue and quarterly employee data for all years, but given that we do not, the most sensible thing to do is to keep things annualized except for the one year of interest, when Mayer arrives as CEO. It's worth doing this because otherwise the large round of layoffs that immediately preceded her arrival would be misattributed to her tenure as CEO. The upshot is that we have two observations for 2012 in the dataset. They have the same revenue data but different employee numbers. The figures can be found in the yahoo dataset.

```
head(yahoo)
```

```
##    Year Revenue Employees Mayer
## 1 2004    3574      7600    No
## 2 2005    5257      9800    No
## 3 2006    6425     11400    No
## 4 2007    6969     14300    No
## 5 2008    7208     13600    No
## 6 2009    6460     13900    No
```

The redrawing is straightforward. We could just draw a scatterplot and color the points by whether Mayer was CEO at the time. By now you should know how to do this quite easily. We can take a small step further by making a scatterplot but also holding on to the temporal element beloved of business analysts. We can use geom_path() and use line segments to "join the dots" of the yearly observations in order, labeling each point with its year. The result (fig. 8.22) is a plot that shows the trajectory of the company over time, like a snail moving across a flagstone. Again, bear in mind that we have two observations for 2012.

```
p ← ggplot(data = yahoo,
            mapping = aes(x = Employees, y = Revenue))
p + geom_path(color = "gray80") +
    geom_text(aes(color = Mayer, label = Year),
              size = 3, fontface = "bold") +
    theme(legend.position = "bottom") +
    labs(color = "Mayer is CEO",
         x = "Employees", y = "Revenue (Millions)",
         title = "Yahoo Employees vs Revenues, 2004-2014") +
  scale_y_continuous(labels = scales::dollar) +
  scale_x_continuous(labels = scales::comma)
```

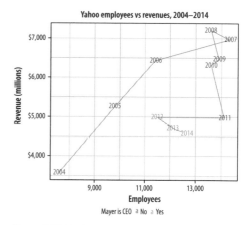

Figure 8.22: Redrawing as a connected scatterplot.

This way of looking at the data suggests that Mayer was appointed after a period of falling revenues and just following a very large round of layoffs, a fairly common pattern with the leadership of large firms. Since then, through either new hires or acquisitions, employee numbers have crept back up a little while revenues have continued to fall. This version more clearly conveys what the original slide was trying to get across.

Alternatively, we can keep the analyst community happy by putting time back on the x-axis and plotting the ratio of revenue to employees on the y-axis. This gives us the linear time-trend back, only in a more sensible fashion (fig. 8.23). We begin the plot by using geom_vline() to add a vertical line marking Mayer's accession to the CEO position.

```
p ← ggplot(data = yahoo,
            mapping = aes(x = Year, y = Revenue/Employees))

p + geom_vline(xintercept = 2012) +
    geom_line(color = "gray60", size = 2) +
    annotate("text", x = 2013, y = 0.44,
             label = " Mayer becomes CEO", size = 2.5) +
    labs(x = "Year\n",
         y = "Revenue/Employees",
         title = "Yahoo Revenue to Employee Ratio, 2004-2014")
```

Figure 8.23: Plotting the ratio of revenue to employees against time. (From the December 2015 Yahoo! Investor Presentation: A Better Plan for Yahoo Shareholders.)

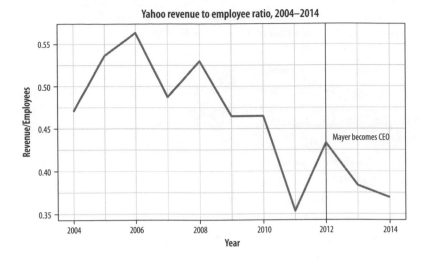

Figure 8.24: Data on the structure of U.S. student debts as of 2016.

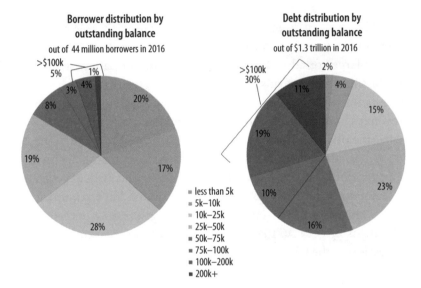

Saying no to pie

For a third example, we turn to pie charts. Figure 8.24 shows a pair of charts from a New York Federal Reserve Bank briefing on the structure of debt in the United States (Chakrabarti et al. 2017). As we saw in chapter 1, the perceptual qualities of pie charts are not great. In a single pie chart, it is usually harder than it should be to estimate and compare the values shown, especially when there are more than a few wedges and when there are a number of wedges reasonably close in size. A Cleveland dotplot or a bar chart is usually a much more straightforward way of comparing quantities. When comparing the wedges between two pie charts, as in this case, the task is made harder again as the viewer has to ping back and forth between the wedges of each pie and the vertically oriented legend underneath.

There is an additional wrinkle in this case. The variable broken down in each pie chart is not only categorical, it is also ordered from low to high. The data describe the percent of all borrowers and the percent of all balances divided up across the size of balances owed, from less than five thousand dollars to more than two hundred thousand dollars. It's one thing to use a pie chart to display shares of an unordered categorical variable, such as percent of total sales due to pizza, lasagna, and risotto. Keeping track of ordered categories in a pie chart is harder again, especially when we want to make a comparison between two distributions. The wedges of these two pie charts *are* ordered (clockwise, from the top), but it's not so easy to follow them. This is partly because of the pie-ness of the chart and partly because the color palette chosen for the categories is not sequential. Instead it is unordered. The colors allow the debt categories to be distinguished but don't pick out the sequence from low to high values.

So not only is a less than ideal plot type being used here, it's being made to do a lot more work than usual, and with the wrong sort of color palette. As is often the case with pie charts, the compromise made to facilitate interpretation is to display all the numerical values for every wedge, and also to add a summary outside the pie. If you find yourself having to do this, it's worth asking whether the chart could be redrawn, or whether you might as well just show a table instead.

Here are two ways we might redraw these pie charts. As usual, neither approach is perfect. Or rather, each approach draws attention to features of the data in slightly different ways. Which works best depends on what parts of the data we want to highlight. The data are in an object called studebt.

```
head(studebt)
```

```
## # A tibble: 6 x 4
##   Debt      type         pct Debtrc
##   <ord>     <fct>      <int> <ord>
## 1 Under $5  Borrowers     20 Under $5
```

Figure 8.25: Faceting the pie charts.

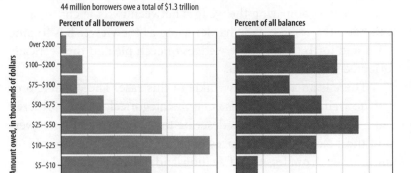

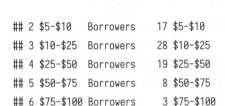

```
##  2 $5-$10    Borrowers    17 $5-$10
##  3 $10-$25   Borrowers    28 $10-$25
##  4 $25-$50   Borrowers    19 $25-$50
##  5 $50-$75   Borrowers     8 $50-$75
##  6 $75-$100  Borrowers     3 $75-$100
```

Our first effort to redraw the pie charts (fig. 8.25) uses a faceted comparison of the two distributions. We set up some labels in advance, as we will reuse them. We also make a special label for the facets.

```r
p_xlab ← "Amount Owed, in thousands of Dollars"
p_title ← "Outstanding Student Loans"
p_subtitle ← "44 million borrowers owe a total of $1.3 trillion"
p_caption ← "Source: FRB NY"


f_labs ← c(`Borrowers` = "Percent of\nall Borrowers",
           `Balances` = "Percent of\nall Balances")


p ← ggplot(data = studebt,
           mapping = aes(x = Debt, y = pct/100, fill = type))
p + geom_bar(stat = "identity") +
    scale_fill_brewer(type = "qual", palette = "Dark2") +
    scale_y_continuous(labels = scales::percent) +
    guides(fill = FALSE) +
    theme(strip.text.x = element_text(face = "bold")) +
    labs(y = NULL, x = p_xlab,
```

```
   caption = p_caption,
   title = p_title,
   subtitle = p_subtitle) +
 facet_grid(~ type, labeller = as_labeller(f_labs)) +
 coord_flip()
```

There is a reasonable amount of customization in this graph. First, the text of the facets is made bold in the theme() call. The graphical element is first named (strip.text.x) and then modified using the element_text() function. We also use a custom palette for the fill mapping, via scale_fill_brewer(). And finally we relabel the facets to something more informative than their bare variable names. This is done using the labeller argument and the as_labeller() function inside the facet_grid() call. At the beginning of the plotting code, we set up an object called f_labs, which is in effect a tiny data frame that associates new labels with the values of the type variable in studebt. We use backticks (the angled quote character located next to the '1' key on U.S. keyboards) to pick out the values we want to relabel. The as_labeller() function takes this object and uses it to create new text for the labels when facet_grid() is called.

Substantively, how is this plot better than the pie charts? We split the data into the two categories and showed the percentage shares as bars. The percent scores are on the x-axis. Instead of using color to distinguish the debt categories, we put their values on the y-axis instead. This means we can compare within a category just by looking down the bars. For instance, the left-hand panel shows that almost a fifth of the 44 million people with student debt owe less than five thousand dollars. Comparisons across categories are now easier as well, as we can scan across a row to see, for instance, that while just 1 percent or so of borrowers owe more than $200,000, that category accounts for more than 10 percent of all student debt.

We could also have made this bar chart by putting the percentages on the y-axis and the categories of amount owed on the x-axis. When the categorical axis labels are long, though, I generally find it's easier to read them on the y-axis. Finally, while it looks nice and helps a little to have the two categories of debt distinguished by color, the colors on the graph are not encoding or mapping any information in the data that is not already taken care of by the faceting. The fill mapping is useful but also redundant.

Outstanding student loans

44 million borrowers owe a total of $1.3 trillion

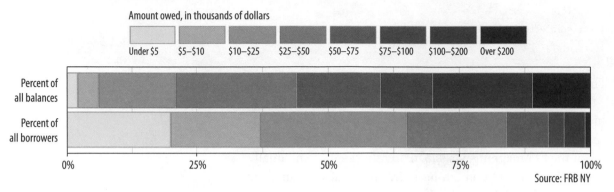

Figure 8.26: Debt distributions as horizontally segmented bars.

This graph could easily be in black and white and would be just as informative.

One thing that is not emphasized in a faceted chart like this is the idea that each of the debt categories is a share or percentage of a total amount. That is what a pie chart emphasizes more than anything, but as we saw there's a perceptual price to pay for that, especially when the categories are ordered. But maybe we can hang on to the emphasis on shares by using a different kind of barplot. Instead of having separate bars distinguished by heights, we can array the percentages for each distribution proportionally within a single bar. We will make a stacked bar chart with just two main bars, (fig. 8.26) and lie them on their side for comparison.

```r
library(viridis)

p ← ggplot(studebt, aes(y = pct/100, x = type, fill = Debtrc))
p + geom_bar(stat = "identity", color = "gray80") +
  scale_x_discrete(labels = as_labeller(f_labs)) +
  scale_y_continuous(labels = scales::percent) +
  scale_fill_viridis(discrete = TRUE) +
  guides(fill = guide_legend(reverse = TRUE,
                             title.position = "top",
                             label.position = "bottom",
                             keywidth = 3,
                             nrow = 1)) +
```

```
labs(x = NULL, y = NULL,
     fill = "Amount Owed, in thousands of dollars",
     caption = p_caption,
     title = p_title,
     subtitle = p_subtitle) +
theme(legend.position = "top",
      axis.text.y = element_text(face = "bold", hjust = 1, size = 12),
      axis.ticks.length = unit(0, "cm"),
      panel.grid.major.y = element_blank()) +
coord_flip()
```

Once again, there is a substantial amount of customization in this chart. I encourage you to peel it back one option at a time to see how it changes. We use the as_labeller() with f_labs again, but in the labels for the x-axis this time. We make a series of adjustments in the theme() call to customize the purely visual elements of the plot, making the y-axis labels larger, right justified, and bold via element_text(), removing the axis tick-marks, and also removing the y-axis grid lines via element_blank().

More substantively, we take a lot of care about color in figure 8.26. First, we set the border colors of the bars to a light gray in geom_bar() to make the bar segments easier to distinguish. Second, we draw on the viridis library again (as we did with our small-multiple maps in chapter 7), using scale_fill_viridis() for the color palette. Third, we are careful to map the income categories in an ascending sequence of colors, and to adjust the key so that the values run from low to high, from left to right, and from yellow to purple. This is done partly by switching the fill mapping from Debt to Debtrc. The categories of the latter are the same as the former, but the sequence of income levels is coded in the order we want. We also show the legend to the reader first by putting it at the top, under the title and subtitle.

The rest of the work is done in the guides() call. We have not used guides() much thus far except to turn off legends that we did not want to display. But here we see its usefulness. We give guides() a series of instructions about the fill mapping: reverse the direction of the color coding; put the legend title above the key; put the labels for the colors below the key; widen the width of the color boxes a little; and place the whole key on a single row.

`reverse = TRUE`
`title.position`
`label.position`
`keywidth`
`nrow`

This is a lot of work, but if you don't do it the plot will be much harder to read. Again, I encourage you to peel away the layers and options in sequence to see how the plot changes. The version in figure 8.26 lets us more easily see how the categories of dollar amounts owed break down as a percentage of all balances, and as a percent of all borrowers. We can also eyeball comparisons between the two types, especially at the far end of each scale. It's easy to see how a tiny percentage of borrowers account for a disproportionately large share of total debt, for example. But even with all this careful work, estimating the size of each individual segment is still not as easy here as it is in figure 8.25, the faceted version. This is because it's harder to estimate sizes when we don't have an anchor point or baseline scale to compare each piece to. (In the faceted plot, that comparison point was the x-axis.) So the size of the "Under 5" segment in the bottom bar is much easier to estimate than the size of the "$10–25" bar, for instance. Our injunction to take care about using stacked bar charts still has a lot of force, even when we try hard to make the best of them.

8.6 Where to Go Next

We have reached the end of our introduction. From here on, you should be in a strong position to forge ahead in two main ways. The first is to become more confident and practiced with your coding. Learning ggplot should encourage you to learn more about the set of tidyverse tools, and then by extension to learn more about R in general. What you choose to pursue will (and should) be driven by your own needs and interests as a scholar or data scientist. The most natural text to look at next is Garrett Grolemund and Hadley Wickham's *R for Data Science* (2016), which introduces tidyverse components that we have drawn on here but not pursued in depth. Other useful texts include Chang (2013) and Roger Peng's *R Programming for Data Science* (2016). The most thorough introduction to ggplot in particular can be found in Wickham (2016).

Pushing ahead to use ggplot for new kinds of graphs will eventually get you to the point where ggplot does not quite do what you need or does not quite provide the sort of geom you want. In that case, the first place to look is the world of extensions to the

r4ds.had.co.nz/

leanpub.com/rprogramming

ggplot framework. Daniel Emaasit's overview of add-on packages for ggplot is the best place to begin your search. We have used a few extensions in the book already. Like ggrepel and ggridges, extensions typically provide a new geom or two to work with, which may be just what you need. Sometimes, as with Thomas Lin Pedersen's `ggraph`, you get a whole family of geoms and associated tools—in `ggraph`'s case, a suite of tidy methods for the visualization of network data. Other modeling and analysis tasks may require more custom work, or coding that is closely connected to the kind of analysis being done. Harrell (2016) provides many clearly worked examples, mostly based on ggplot; Gelman & Hill (2018) and Imai (2017) also introduce contemporary methods using R; Silge & Robinson (2017) present a tidy approach to analyzing and visualizing textual data; while Friendly & Meyer (2017) thoroughly explore the analysis of discrete data, an area that is often challenging to approach visually.

ggplot2-exts.org

The second way you should push ahead is by looking at and thinking about other people's graphs. The R Graph Gallery, run by Yan Holtz, is a useful collection of examples of many kinds of graphics drawn with ggplot and other R tools. PolicyViz, a site run by Jon Schwabish, covers a range of topics on data visualization. It regularly features case studies where visualizations are reworked to improve them or cast new light on the data they present. But do not just look for examples that have code with them to begin with. As I have said before, a real strength of ggplot is the grammar of graphic that underpins it. That grammar is a model you can use to look at and interpret *any* graph, no matter how it was produced. It gives you a vocabulary that lets you say what the data, mappings, geoms, scales, guides, and layers of any particular graph might be. And because the grammar is implemented as the ggplot library, it is a short step from being able to anatomize the structure of a graph to being able to sketch an outline of the code you could write to reproduce it yourself.

r-graph-gallery.com

policyviz.com

While its underlying principles and goals are relatively stable, the techniques and tools of research are changing. This is especially true within the social sciences (Salganik 2018). Data visualization is an excellent entry point to these new developments. Our tools for it are more versatile and powerful than ever. You should look at your data. Looking is not a replacement for thinking. It cannot force you to be honest; it cannot magically prevent you from making mistakes; and it cannot make your ideas true. But if you analyze data, visualization can help you uncover features in it. If you are honest, it can help you live up to your own standards. When you inevitably make errors, it can help you find and correct them. And if you have an idea and some good evidence for it, it can help you show it in a compelling way.

Acknowledgments

Software is a kind of condensed collaboration. It's a big help in everyday life but is also easy to take for granted. The research and teaching I do depend on tools that other people freely write, maintain, and distribute. My first debt is therefore to all those who produce and maintain R, the infrastructure that supports it, and the packages built on its foundations. The particular libraries used in the book are cited throughout the text and listed in the references. I am also grateful to those in the R community who helped me while I wrote this book, whether directly, through comments and suggestions; indirectly, by independently solving problems I ran into myself; or unwittingly, via the excellent example of their own open and generous style of work. In particular I thank Jenny Bryan, Mine Çetinkaya-Rundel, Dirk Eddelbuettel, Thomas Lumley, John MacFarlane, Bob Rudis, Hadley Wickham, Claus O. Wilke, and Yihui Xie.

A conversation with Chris Bail, a coauthored paper with Jim Moody (Healy & Moody 2014), and a suggestion from Steve Vaisey got me started on this project. Duke University and the Kenan Institute for Ethics gave me time to see it through. Martin Ruef and Suzanne Shanahan made the necessary room. My students and seminar participants at Duke, Yale, McGill, the University of Oslo, and Statistical Horizons were test pilots for much of the material and provided invaluable feedback. I thank Andy Papachristos, Tom Lyttleton, Torkild Lyngstad, Amélie Quesnel-Vallée, Shelley Clark, and Paul Allison for the external teaching opportunities.

At Princeton University Press, Meagan Levinson has been an ideal editor in every respect. Her expert guidance and enthusiasm throughout the writing and production of the book made everything move a lot quicker than expected. Four anonymous readers for the Press provided detailed and helpful comments that improved the manuscript substantially. I also received excellent feedback from Andrew Gelman, Eszter Hargittai, Matissa Hollister, Eric Lawrence, and Matt Salganik. Any remaining errors are of course my own.

For many years, Laurie Paul gave her encouragement and support to this and other projects. I thank her for that. *Is cuma leis an mhaidin cad air a ngealann sí.*

For their professionalism as colleagues or their kindness as friends, I am grateful to Sam Anthony, Courtney Bender, Andrea Deeker, Mary Dixon-Woods, John Evans, Tina Fetner, Pierre-Olivier Gourinchas, Erin Kelly, Liz Morgan, Brian Steensland, Craig Upright, Rebekah Estrada Vaisey, and Steve Vaisey. Marion Fourcade has both these qualities but also a third—patience as a coauthor—that I fear I have tried beyond reason.

Much of this book was written on the Robertson Scholars Bus, which goes back and forth between Duke and Chapel Hill on the half hour during term. It's a big help in everyday life but is also easy to take for granted. Its existence is worth any amount of hand waving about connection and collaboration. The best seats are near the front, facing sideways.

Appendix

This appendix contains supplemental information about various aspects of R and ggplot that you are likely to run into as you use it. You are at the beginning of a process of discovering practical problems that are an inevitable part of using software. This is often frustrating. But feeling stumped is a standard experience for everyone who writes code. Each time you figure out the solution to your problem, you acquire more knowledge about how and why things go wrong, and more confidence about how to tackle the next glitch that comes along.

1 A little more about R

How to read an R help page

Functions, datasets, and other built-in objects in R are documented in its help system. You can search or browse this documentation via the "Help" tab in RStudio's lower right-hand window. The quality of R's help pages varies somewhat. They tend to be on the terse side. However, they all have essentially the same structure, and it is useful to know how to read them. Figure A.1 provides an overview of what to look for. Remember, functions take inputs, perform actions, and return outputs. Something goes in, it gets worked on, and then something comes out. That means you want to know what the function *requires*, what it *does*, and what it *returns*. What it requires is shown in the *Usage* and *Arguments* sections of the help page. The names of the required and optional arguments are given in the order the function expects them. Some arguments have default values. In the case of the mean() function, the argument na.rm is set to FALSE by default. These will be shown in the *Usage* section. If a named argument has no default, you will have to give it a value. Depending on what the argument is, this might be a logical value, a number, a dataset, or any other object.

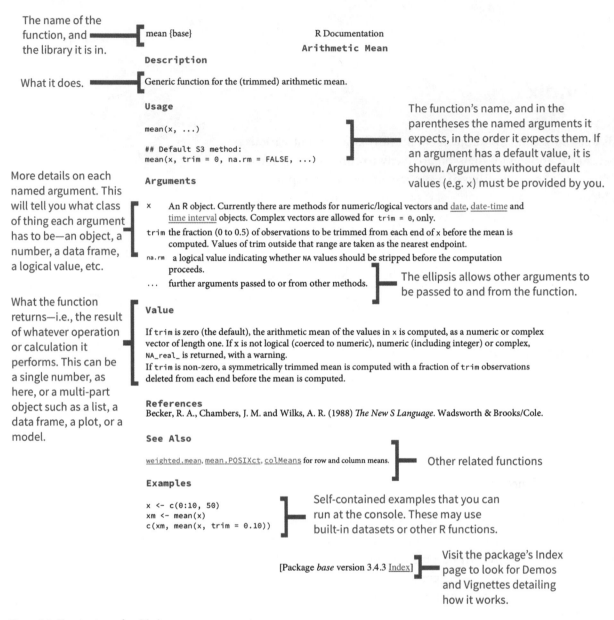

The name of the function, and the library it is in.

mean {base}   R Documentation
                                                        Arithmetic Mean

Description

What it does.

Generic function for the (trimmed) arithmetic mean.

Usage

The function's name, and in the parentheses the named arguments it expects, in the order it expects them. If an argument has a default value, it is shown. Arguments without default values (e.g. x) must be provided by you.

```
mean(x, ...)

## Default S3 method:
mean(x, trim = 0, na.rm = FALSE, ...)
```

Arguments

More details on each named argument. This will tell you what class of thing each argument has to be—an object, a number, a data frame, a logical value, etc.

x   An R object. Currently there are methods for numeric/logical vectors and date, date-time and time interval objects. Complex vectors are allowed for trim = 0, only.

trim   the fraction (0 to 0.5) of observations to be trimmed from each end of x before the mean is computed. Values of trim outside that range are taken as the nearest endpoint.

na.rm   a logical value indicating whether NA values should be stripped before the computation proceeds.

...   further arguments passed to or from other methods.

The ellipsis allows other arguments to be passed to and from the function.

Value

What the function returns—i.e., the result of whatever operation or calculation it performs. This can be a single number, as here, or a multi-part object such as a list, a data frame, a plot, or a model.

If trim is zero (the default), the arithmetic mean of the values in x is computed, as a numeric or complex vector of length one. If x is not logical (coerced to numeric), numeric (including integer) or complex, NA_real_ is returned, with a warning.
If trim is non-zero, a symmetrically trimmed mean is computed with a fraction of trim observations deleted from each end before the mean is computed.

References
Becker, R. A., Chambers, J. M. and Wilks, A. R. (1988) *The New S Language.* Wadsworth & Brooks/Cole.

See Also

weighted.mean, mean.POSIXct, colMeans for row and column means.

Other related functions

Examples

```
x <- c(0:10, 50)
xm <- mean(x)
c(xm, mean(x, trim = 0.10))
```

Self-contained examples that you can run at the console. These may use built-in datasets or other R functions.

[Package *base* version 3.4.3 Index]

Visit the package's Index page to look for Demos and Vignettes detailing how it works.

Figure A.1: The structure of an R help page.

The other part to look at closely is the *Value* section, which tells you what the function returns once it has done its calculation. Again, depending on what the function is, this might simply be a single number or other short bit of output. But it could also be something as complex as a ggplot figure or a model object consisting of many separate parts organized as a list.

Well-documented packages will often have *demos* and *vignettes* attached to them. These are meant to describe the package as a whole, rather than specific functions. A good package vignette will often have one or more fully worked examples together with a discussion describing how the package works and what it can do. To see if there are any package vignettes, click the link at the bottom of the function's help page to be taken to the package index. Any available demos, vignettes, or other general help will be listed at the top.

The basics of accessing and selecting things

Generally speaking, the tidyverse's preferred methods for data subsetting, filtering, slicing and selecting will keep you away from the underlying mechanics of selecting and extracting elements of vectors, matrices, or tables of data. Carrying out these operations through functions like select(), filter(), subset(), and merge() is generally safer and more reliable than accessing elements directly. However, it is worth knowing the basics of these operations. Sometimes accessing elements directly is the most convenient thing to do. More important, we may use these techniques in small ways in our code with some regularity. Here we very briefly introduce some of R's selection operators for vectors, arrays, and tables.

Consider the my_numbers and your_numbers vectors again.

```
my_numbers ← c(1, 2, 3, 1, 3, 5, 25)
your_numbers ← c(5, 31, 71, 1, 3, 21, 6)
```

To access any particular element in my_numbers, we use square brackets. Square brackets are not like the parentheses after functions. They are used to pick out an element indexed by its position:

```
my_numbers[4]
```

```
## [1] 1
```

```
my_numbers[7]
```

```
## [1] 25
```

Putting the number *n* inside the brackets will give us (or "return") the *n*th element in the vector, assuming there is one. To access a *sequence* of elements within a vector we can do this:

```
my_numbers[2:4]
```

```
## [1] 2 3 1
```

This shorthand notation tells R to count from the second to the fourth element, inclusive. We are not restricted to selecting contiguous elements either. We can make use of our c() function again:

```
my_numbers[c(2, 4)]
```

```
## [1] 2 1
```

R evaluates the expression c(2,4) first and then extracts just the second and the fourth element from my_numbers, ignoring the others. You might wonder why we didn't just write my_numbers[2,3] directly. The answer is that this notation is used for objects arrayed in two dimensions (i.e., something with rows and columns), like matrices, data frames, or tibbles. We can make a two-dimensional object by creating two different vectors with the c() function and using the tibble() function to collect them together:

```
my_tb ← tibble(mine = c(1, 4, 5, 8:11), yours = c(3, 20, 16,
    34:31))

class(my_tb)
```

```
## [1] "tbl_df"     "tbl"         "data.frame"
```

```
my_tb
```

```
## # A tibble: 7 x 2
##     mine yours
##    <dbl> <dbl>
## 1    1.    3.
## 2    4.   20.
## 3    5.   16.
## 4    8.   34.
```

```
## 5    9.   33.
## 6    10.  32.
## 7    11.  31.
```

We index data frames, tibbles, and other arrays by row first, and then by column. Arrays may also have more than two dimensions.

```
my_tb[3, 1]  # Row 3 Col 1
```

```
## # A tibble: 1 x 1
##    mine
##    <dbl>
## 1   5.
```

```
my_tb[1, 2]  # Row 1, Col 2
## # A tibble: 1 x 1
##    yours
##    <dbl>
## 1   3.
```

The columns in our tibble have names. We can select elements using them, too. We do this by putting the name of the column in quotes where we previously put the index number of the column:

```
my_tb[3, "mine"]  # Row 3 Col 1
```

```
## # A tibble: 1 x 1
##     mine
##    <dbl>
## 1    5.
```

```
my_tb[1, "yours"]  # Row 1, Col 2
```

```
## # A tibble: 1 x 1
##    yours
##    <dbl>
## 1    3.
```

```
my_tb[3, "mine"]  # Row 3 Col 1
```

In these chunks of code you will see some explanatory text set off by the hash symbol, #. In R's syntax, the hash symbol is used to designate a comment. On any line of code, text that appears after a # symbol will be ignored by R's interpreter. It won't be evaluated and it won't trigger a syntax error.

```
## # A tibble: 1 x 1
##    mine
##   <dbl>
## 1   5.
```

```
my_tb[1, "yours"]  # Row 1, Col 2
```

```
## # A tibble: 1 x 1
##   yours
##   <dbl>
## 1   3.
```

If we want to get all the elements of a particular column, we can leave out the row index. This will mean all the rows will be included for whichever column we select.

```
my_tb[, "mine"]  # All rows, Col 1
```

```
## # A tibble: 7 x 1
##    mine
##   <dbl>
## 1   1.
## 2   4.
## 3   5.
## 4   8.
## 5   9.
## 6  10.
## 7  11.
```

We can do this the other way around, too, selecting a particular row and showing all columns:

```
my_tb[4, ]  # Row 4, all cols
```

```
## # A tibble: 1 x 2
##    mine yours
##   <dbl> <dbl>
## 1   8.   34.
```

A better way of accessing particular columns in a data frame is via the $ operator, which can be used to extract components of various sorts of object. This way we append the name of the column we want to the name of the object it belongs to:

```
my_tb$mine
```

```
## [1]  1  4  5  8  9 10 11
```

Elements of many other objects can be extracted in this way, too, including nested objects.

```
out ← lm(mine ~ yours, data = my_tb)
```

```
out$coefficients
```

```
## (Intercept)      yours
## -0.0801192   0.2873422
```

```
out$call
## lm(formula = mine ~ yours, data = my_tb)
```

```
out$qr$rank   # nested
```

```
## [1] 2
```

Finally, in the case of data frames, the $ operator also lets us add new columns to the object. For example, we can add the first two columns together, row by row. To create a column in this way, we put the $ and the name of the new column on the left side of the assignment.

```
my_tb$ours ← my_tb$mine + my_tb$yours
my_tb
```

```
## # A tibble: 7 x 3
##    mine yours  ours
##   <dbl> <dbl> <dbl>
## 1   1.    3.    4.
## 2   4.   20.   24.
## 3   5.   16.   21.
## 4   8.   34.   42.
## 5   9.   33.   42.
## 6  10.   32.   42.
## 7  11.   31.   42.
```

In this book we do not generally access data via [or $. It is particularly bad practice to access elements by their index number only, as opposed to using names. In both cases, and especially the latter, it is too easy to make a mistake and choose the wrong columns or rows. In addition, if our table changes shape later on (e.g., due to the addition of new original data), then any absolute reference to the position of columns (rather than to their names) is very likely to break. Still, we do use the c() function for small tasks quite regularly, so it's worth understanding how it can be used to pick out elements from vectors.

TABLE A.1
Some untidy data.

name	treatmenta	treatmentb
John Smith	NA	18
Jane Doe	4	1
Mary Johnson	6	7

TABLE A.2
The same data, still untidy, but in a different way.

treatment	John Smith	Jane Doe	Mary Johnson
a	NA	4	6
b	18	1	7

TABLE A.3
Tidied data. Every variable a column, every observation a row.

name	treatment	n
Jane Doe	a	4
Jane Doe	b	1
John Smith	a	NA
John Smith	b	18
Mary Johnson	a	6
Mary Johnson	b	7

Tidy data

Working with R and ggplot is much easier if the data you use is in the right shape. Ggplot wants your data to be *tidy*. For a more thorough introduction to the idea of tidy data, see chapters 5 and 12 of Wickham & Grolemund (2016). To get a sense of what a tidy dataset looks like in R, we will follow the discussion in Wickham (2014). In a tidy dataset,

1. *Each variable is a column.*
2. *Each observation is a row.*
3. *Each type of observational unit forms a table.*

For most of your data analysis, the first two points are the most important. The third point might be a little unfamiliar. It is a feature of "normalized" data from the world of databases, where the goal is to represent data in a series of related tables with minimal duplication (Codd 1990). Data analysis more usually works with a single large table of data, often with considerable duplication of some variables down the rows.

Data presented in summary tables is often not "tidy" as defined here. When structuring our data, we need to be clear about how our data is arranged. If your data is not tidily arranged, the chances are good that you will have more difficulty, and maybe a *lot* more difficulty, getting ggplot to draw the figure you want.

For example, consider table A.1 and table A.2, from Wickham's discussion. They present the same data in different ways, but each would cause trouble if we tried to work with it in ggplot to make a graph. Table A.3 shows the same data once again, this time in a tidied form.

Table A-1. Years of school completed by people 25 years and over, by age and sex: selected years 1940 to 2016
(Numbers in thousands. Noninstitutionalized population except where otherwise specified.)

Age, sex, and years	Total	Years of school completed						Median
		Elementary		High school		College		
		0 to 4 years	5 to 8 years	1 to 3 years	4 years	1 to 3 years	4 years or more	
25 years and older								
Male								
2016	103,372	1,183	3,513	7,144	30,780	26,468	34,283	(NA)
2015	101,887	1,243	3,669	7,278	30,997	25,778	32,923	(NA)
2014	100,592	1,184	3,761	7,403	30,718	25,430	32,095	(NA)
2013	99,305	1,127	3,836	7,314	30,014	25,283	31,731	(NA)
2012	98,119	1,231	3,879	7,388	30,216	24,632	30,766	(NA)
2011	97,220	1,234	3,883	7,443	30,370	24,319	29,971	(NA)
2010	96,325	1,279	3,931	7,705	30,682	23,570	29,158	(NA)
2009	95,518	1,372	4,027	7,754	30,025	23,634	28,706	(NA)
2008	94,470	1,310	4,136	7,853	29,491	23,247	28,433	(NA)
2007	93,421	1,458	4,249	8,294	29,604	22,219	27,596	(NA)
2006	92,233	1,472	4,395	7,940	29,380	22,136	26,910	(NA)
2005	90,899	1,505	4,402	7,787	29,151	21,794	26,259	(NA)

Figure A.2: Untidy data from the census.

Hadley Wickham notes five main ways tables of data tend not to be tidy:

1. Column headers are values, not variable names.
2. Multiple variables are stored in one column.
3. Variables are stored in both rows and columns.
4. Multiple types of observational units are stored in the same table.
5. A single observational unit is stored in multiple tables.

Data comes in an untidy form all the time, often for the good reason that it can be presented that way using much less space, or with far less repetition of labels and row elements. Figure A.2 shows a the first few rows of a table of U.S. Census Bureau data about educational attainment in the United States. To begin with, it's organized as a series of subtables down the spreadsheet, broken out by age and sex. Second, the underlying variable of interest, "Years of School Completed," is stored across several columns, with an additional variable (level of schooling) included across the columns also. It is not hard to get the table into a slightly more regular format by eliminating the blank rows and explicitly naming the subtable rows. One can do this manually and get to the point where it can be read in as an Excel or CSV file. This is not ideal,

readxl.tidyverse.org

as manually cleaning data runs against the commitment to do as much as possible programmatically. We can automate the process somewhat. The tidyverse comes with a `readxl` package that tries to ease the pain.

```
## # A tibble: 366 x 11
##    age   sex    year total elem4 elem8   hs3   hs4 coll3 coll4 median
##    <chr> <chr> <int> <int> <int> <int> <dbl> <dbl> <dbl> <dbl>  <dbl>
##  1 25-34 Male   2016 21845   116   468 1427. 6386. 6015. 7432.     NA
##  2 25-34 Male   2015 21427   166   488 1584. 6198. 5920. 7071.     NA
##  3 25-34 Male   2014 21217   151   512 1611. 6323. 5910. 6710.     NA
##  4 25-34 Male   2013 20816   161   582 1747. 6058. 5749. 6519.     NA
##  5 25-34 Male   2012 20464   161   579 1707. 6127. 5619. 6270.     NA
##  6 25-34 Male   2011 20985   190   657 1791. 6444. 5750. 6151.     NA
##  7 25-34 Male   2010 20689   186   641 1866. 6458. 5587. 5951.     NA
##  8 25-34 Male   2009 20440   184   695 1806. 6495. 5508. 5752.     NA
##  9 25-34 Male   2008 20210   172   714 1874. 6356. 5277. 5816.     NA
## 10 25-34 Male   2007 20024   246   757 1930. 6361. 5137. 5593.     NA
## # ... with 356 more rows
```

The tidyverse has several tools to help you get the rest of the way in converting your data from an untidy to a tidy state. These can mostly be found in the `tidyr` and `dplyr` packages. The former provides functions for converting, for example, wide-format data to long-format data, as well as assisting with the business of splitting and combining variables that are untidily stored. The latter has tools that allow tidy tables to be further filtered, sliced, and analyzed at different grouping levels, as we have seen throughout this book.

With our `edu` object, we can use the `gather()` function to transform the schooling variables into a *key-value* arrangement. The key is the underlying variable, and the value is the value it takes for that observation. We create a new object, `edu_t` in this way.

```
edu_t ← gather(data = edu,
               key = school,
               value = freq,
               elem4:coll4)


head(edu_t)
```

```
## # A tibble: 6 x 7
##   age   sex    year total median school  freq
##   <chr> <chr> <int> <int>  <dbl> <chr>  <dbl>
## 1 25-34 Male   2016 21845     NA elem4   116.
## 2 25-34 Male   2015 21427     NA elem4   166.
## 3 25-34 Male   2014 21217     NA elem4   151.
## 4 25-34 Male   2013 20816     NA elem4   161.
## 5 25-34 Male   2012 20464     NA elem4   161.
## 6 25-34 Male   2011 20985     NA elem4   190.
```

```
tail(edu_t)
```

```
## # A tibble: 6 x 7
##   age   sex     year total median school  freq
##   <chr> <chr>  <int> <int>  <dbl> <chr>  <dbl>
## 1 55>   Female  1959 16263   8.30 coll4   688.
## 2 55>   Female  1957 15581   8.20 coll4   630.
## 3 55>   Female  1952 13662   7.90 coll4   628.
## 4 55>   Female  1950 13150   8.40 coll4   436.
## 5 55>   Female  1947 11810   7.60 coll4   343.
## 6 55>   Female  1940  9777   8.30 coll4   219.
```

The educational categories previously spread over the columns have been gathered into two new columns. The school variable is the *key* column. It contains all the education categories that were previously given across the column headers, from zero to four years of elementary school to four or more years of college. They are now stacked up on top of each other in the rows. The freq variable is the *value* column and contains the unique value of schooling for each level of that variable. Once our data is in this long-form shape, it is ready for easy use with ggplot and related tidyverse tools.

2 Common Problems Reading in Data

Date formats

Date formats can be annoying. First, times and dates must be treated differently from ordinary numbers. Second, there are many different date formats, differing both in the precision with which

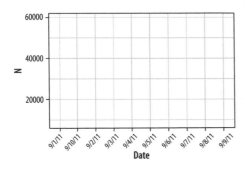

Figure A.3: A bad date.

they are stored and in the convention they follow about how to display years, months, days, and so on. Consider the following data:

```
head(bad_date)
```

```
## # A tibble: 6 x 2
##   date       N
##   <chr>    <int>
## 1 9/1/11  44426
## 2 9/2/11  55112
## 3 9/3/11  19263
## 4 9/4/11  12330
## 5 9/5/11   8534
## 6 9/6/11  59490
```

The data in the date column has been read in as a character string, but we want R to treat it as a date. If can't treat it as a date, we get bad results (fig. A.3).

```
p <- ggplot(data = bad_date, aes(x = date, y = N))
p + geom_line()
```

```
## geom_path: Each group consists of only one observation.
## Do you need to adjust the group aesthetic?
```

What has happened? The problem is that ggplot doesn't know date consists of dates. As a result, when we ask to plot it on the x-axis, it tries to treat the unique elements of date like a categorical variable instead. (That is, as a factor.) But because each date is unique, its default effort at grouping the data results in every group having only one observation in it (i.e., that particular row). The ggplot function knows something is odd about this and tries to let you know. It wonders whether we've failed to set group = <something> in our mapping.

For the sake of it, let's see what happens when the bad date values are *not* unique. We will make a new data frame by stacking two copies of the data on top of each other. The rbind() function does this for us. We end up in figure A.4 with two copies of every observation.

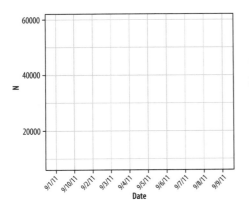

Figure A.4: Still bad.

```
bad_date2 ← rbind(bad_date, bad_date)

p ← ggplot(data = bad_date2, aes(x = date, y = N))
p + geom_line()
```

Now ggplot doesn't complain at all, because there's more than one observation per (inferred) group. But the plot is still wrong!

We will fix this problem using the lubridate package. It provides a suite of convenience functions for converting date strings in various formats and with various separators (such as / or -) into objects of class Date that R knows about. Here our bad dates are in a month/day/year format, so we use mdy(). Consult the lubridate package's documentation to learn more about similar convenience functions for converting character strings where the date components appear in a different order.

```
# install.packages('lubridate')
library(lubridate)

bad_date$date ← mdy(bad_date$date)
head(bad_date)
```

```
## # A tibble: 6 x 2
##    date            N
##    <date>      <int>
## 1 2011-09-01  44426
## 2 2011-09-02  55112
## 3 2011-09-03  19263
## 4 2011-09-04  12330
## 5 2011-09-05   8534
## 6 2011-09-06  59490
```

Now date has a Date class. Let's try the plot again (fig. A.5).

```
p ← ggplot(data = bad_date, aes(x = date, y = N))
p + geom_line()
```

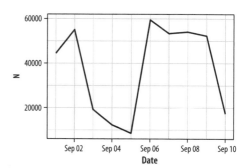

Figure A.5: Much better.

Year-only dates

Many variables are measured by the year and supplied in the data as a four-digit number rather than as a date. This can sometimes cause headaches when we want to plot year on the x-axis. It happens most often when the time series is relatively short. Consider this data:

```
url ← "https://cdn.rawgit.com/kjhealy/viz-organdata/master/organdonation.csv"

bad_year ← read_csv(url)
bad_year %>% select(1:3) %>% sample_n(10)
```

```
## # A tibble: 10 x 3
##    country          year donors
##    <chr>           <int>  <dbl>
##  1 United States    1994   19.4
##  2 Australia        1999   8.67
##  3 Canada           2001   13.5
##  4 Australia        1994   10.2
##  5 Sweden           1993   15.2
##  6 Ireland          1992   19.5
##  7 Switzerland      1997   14.3
##  8 Ireland          2000   17.6
##  9 Switzerland      1998   15.4
## 10 Norway             NA     NA
```

This is a version of organdata but in a less clean format. The year variable is an integer (its class is <int>) and not a date. Let's say we want to plot donation rate against year.

```
p ← ggplot(data = bad_year, aes(x = year, y = donors))
p + geom_point()
```

The decimal point on the x-axis labels in figure A.6 is unwanted. We could sort this out cosmetically, by giving scale_x_continuous() a set of breaks and labels that represent the years as characters. Alternatively, we can change the class of the year variable. For convenience, we will tell R that the year variable

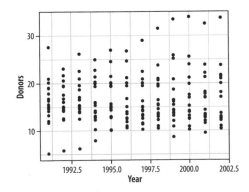

Figure A.6: Integer year shown with a decimal point.

should be treated as a date measure and not an integer. We'll use a home-cooked function, int_to_year(), that takes integers and converts them to dates.

```
bad_year$year ← int_to_year(bad_year$year)
bad_year %>% select(1:3)
```

```
## # A tibble: 238 x 3
##    country   year        donors
##    <chr>     <date>      <dbl>
##  1 Australia NA          NA
##  2 Australia 1991-01-01  12.1
##  3 Australia 1992-01-01  12.4
##  4 Australia 1993-01-01  12.5
##  5 Australia 1994-01-01  10.2
##  6 Australia 1995-01-01  10.2
##  7 Australia 1996-01-01  10.6
##  8 Australia 1997-01-01  10.3
##  9 Australia 1998-01-01  10.5
## 10 Australia 1999-01-01   8.67
## # ... with 228 more rows
```

In the process, today's day and month are introduced into the year data, but that is irrelevant in this case, given that our data is observed only in a yearly window to begin with. However, if you wish to specify a generic day and month for all the observations, the function allows you to do this.

Write functions for repetitive tasks

If you are working with a dataset that you will be making a lot of similar plots from or will need to periodically look at in a way that is repetitive but can't be carried out in a single step once and for all, then the chances are that you will start accumulating sequences of code that you find yourself using repeatedly. When this happens, the temptation will be to start copying and pasting these sequences from one analysis to the next. We can see something of this tendency in the code samples for this book. To make the exposition clearer, we have periodically repeated chunks of code that differ only in the dependent or independent variable being plotted.

Try to avoid copying and pasting code repeatedly in this way. Instead, this is an opportunity to write a function to help you out a little. Almost everything in R is accomplished through functions, and it's not difficult to write your own. This is especially the case when you begin by thinking of functions as a way to help you automate some local or smaller task rather than a means of accomplishing some very complex task. R has the resources to help you build complex functions and function libraries, like ggplot itself. But we can start quite small, with functions that help us manage a particular dataset or data analysis.

Remember, functions take *inputs*, perform *actions*, and return *outputs*. For example, imagine a function that adds two numbers, x and y. In use, it might look like this:

```
add_xy(x = 1, y = 7)
```

```
## [1] 8
```

How do we *create* this function? Remember, everything is an object, so functions are just special kinds of object. And everything in R is done via functions. So if we want to make a new function, we will use an existing function to do it. In R, functions are created with `function()`:

```
add_xy ← function(x, y) {
    x + y
}
```

You can see that `function()` is a little different from ordinary functions in two ways. First, the arguments we give it (here, x and y) are for the `add_xy` function that we are *creating*. Second, immediately after the `function(x, y)` statement there's an opening brace, {, followed by a bit of R code that adds x and y, and then the closing brace }. That's the content of the function. We assign this code to the `add_xy` object, and now we have a function that adds two numbers together and returns the result. The `x + y` line inside the parentheses is evaluated as if it were typed at the console, assuming you have told it what x and y are.

```
add_xy(x = 5, y = 2)
```

```
## [1] 7
```

Functions can take many kinds of arguments, and we can also tell them what the default value of each argument should be by specifying it inside the function(...) section. Functions are little programs that have all the power of R at their disposal, including standard things like flow-control through if ... else statements and so on. Here, for instance, is a function that will make a scatterplot for any section in the ASA data, or optionally fit a smoother to the data and plot that instead. Defining a function looks a little like calling one, except that we spell out the steps inside. We also specify the default arguments.

```
plot_section ← function(section="Culture", x = "Year",
                        y = "Members", data = asasec,
                        smooth=FALSE){
    require(ggplot2)
    require(splines)
    # Note use of aes_string() rather than aes()
    p ← ggplot(subset(data, Sname==section),
          mapping = aes_string(x=x, y=y))

    if(smooth == TRUE) {
        p0 ← p + geom_smooth(color = "#999999",
                             size = 1.2, method = "lm",
                             formula = y ~ ns(x, 3)) +
            scale_x_continuous(breaks = c(seq(2005, 2015, 4))) +
            labs(title = section)
    } else {
    p0 ← p + geom_line(color= "#E69F00", size=1.2) +
        scale_x_continuous(breaks = c(seq(2005, 2015, 4))) +
        labs(title = section)
    }

    print(p0)
}
```

This function is not very general. Nor is it particularly robust. But for the use we want to put it to (fig. A.7) it works just fine.

Figure A.7: Using a function to plot your results.

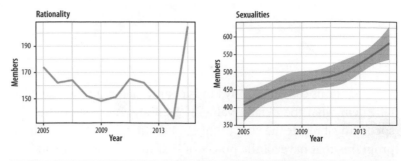

```
plot_section("Rationality")
plot_section("Sexualities", smooth = TRUE)
```

If we were going to work with this data for long enough, we could make the function progressively more general. For example, we can add the special ... argument (which means, roughly, "and any other named arguments") to allow us to pass arguments through to the geom_smooth() function (fig. A.8) in the way we'd expect if we were using it directly. With that in place, we can pick the smoothing method we want.

```
plot_section ← function(section="Culture", x = "Year",
                        y = "Members", data = asasec,
                        smooth=FALSE, ...){
    require(ggplot2)
    require(splines)
    # Note use of aes_string() rather than aes()
    p ← ggplot(subset(data, Sname==section),
          mapping = aes_string(x=x, y=y))


    if(smooth == TRUE) {
        p0 ← p + geom_smooth(color = "#999999",
                            size = 1.2, ...) +
            scale_x_continuous(breaks = c(seq(2005, 2015, 4))) +
            labs(title = section)
    } else {
    p0 ← p + geom_line(color= "#E69F00", size=1.2) +
        scale_x_continuous(breaks = c(seq(2005, 2015, 4))) +
        labs(title = section)
    }


    print(p0)
}
```

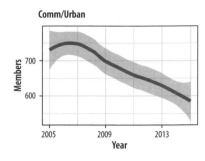 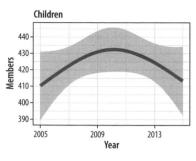

Figure A.8: Our custom function can now pass arguments along to fit different smoothers to section membership data.

```
plot_section("Comm/Urban",
             smooth = TRUE,
             method = "loess")
plot_section("Children",
             smooth = TRUE,
             method = "lm",
             formula = y ~ ns(x, 2))
```

3 Managing Projects and Files

RMarkdown and knitr

Markdown is a loosely standardized way of writing plain text that includes information about the formatting of your document. It was originally developed by John Gruber, with input from Aaron Swartz. The aim was to make a simple format that could incorporate some structural information about the document (such as headings and subheadings, *emphasis*, hyperlinks, lists, and footnotes with minimal loss of readability in plain-text form. A plain-text format like HTML is much more extensive and well defined than Markdown, but Markdown was meant to be simple. Over the years, and despite various weaknesses, it has become a de facto standard. Text editors and note-taking applications support it, and tools exist to convert Markdown not just into HTML (its original target output format) but into many other document types as well. The most powerful of these is Pandoc, which can get you from Markdown to many other formats (and vice versa). Pandoc is what powers RStudio's ability to convert your notes to HTML, Microsoft Word, and PDF documents.

en.wikipedia.org/wiki/Markdown

pandoc.org

rmarkdown.rstudio.com

yihui.name/knitr

Chapter 1 encouraged you to take notes and organize your analysis using RMarkdown and (behind the scenes) knitr. These are R libraries that RStudio makes easy to use. RMarkdown extends Markdown by letting you intersperse your notes with chunks of R code. Code chunks can have labels and a few options that determine how they will behave when the file is processed. After writing your notes and your code, you knit the document (Xie 2015). That is, you feed your .Rmd file to R, which processes the code chunks and produces a new .md where the code chunks have been replaced by their output. You can then turn that Markdown file into a more readable PDF or HTML document, or the Word document that a journal demands you send them.

Behind the scenes in RStudio, this is all done using the knitr and rmarkdown libraries. The latter provides a render() function that takes you from .Rmd to HTML or PDF in a single step. Conversely, if you just want to extract the code you've written from the surrounding text, then you "tangle" the file, which results in an .R file. The strength of this approach is that is makes it much easier to document your work properly. There is just one file for both the data analysis and the writeup. The output of the analysis is created on the fly, and the code to do it is embedded in the paper. If you need to do multiple but identical (or very similar) analyses of different bits of data, RMarkdown and knitr can make generating consistent and reliable reports much easier.

Pandoc's flavor of Markdown is the one used in knitr and RStudio. It allows for a wide range of markup and can handle many of the nuts and bolts of scholarly writing, such as complex tables, citations, bibliographies, references, and mathematics. In addition to being able to produce documents in various *file* formats, it can also produce many different *kinds* of document, from articles and handouts to websites and slide decks. RStudio's RMarkdown website has extensive documentation and examples on the ins and outs of RMarkdown's capabilities, including information on customizing it if you wish.

Writing your notes and papers in a plain-text format like this has many advantages. It keeps your writing, your code, and your results closer together and allows you to use powerful version control methods to keep track of your work and your results. Errors in data analysis often well up out of the gap that typically exists between the procedure used to produce a figure or table in a paper and the subsequent use of that output later. In the ordinary way of doing things, you have the code for your data analysis in one file, the output it produced in another, and the text of your paper in a third file. You do the analysis, collect the output, and copy the relevant results into your paper, often manually reformatting them on the way. Each of these transitions introduces the opportunity for error. In particular, it is easy for a table of results to get detached from the sequence of steps that produced it. Almost everyone who has written a quantitative paper has been confronted with the problem of reading an old draft containing results or figures that need to be revisited or reproduced (as a result of peer review, say) but which lack any information about the circumstances of their creation. Academic papers take a long time to get through the cycle of writing, review, revision, and publication, even when you're working hard the whole time. It is not uncommon to have to return to something you did two years previously in order to answer some question or other from a reviewer. You do not want to have to do everything over from scratch in order to get the right answer. Whatever the challenges of replicating the results of someone else's quantitative analysis, after a fairly short period of time authors themselves find it hard to replicate their *own* work. *Bit-rot* is the term of art in computer science for the seemingly inevitable process of decay that overtakes a project just because you left it alone on your computer for six months or more.

For small and medium-sized projects, plain-text approaches that rely on RMarkdown documents and the tools described here work well. Things become a little more complicated as projects get larger. (This is not an intrinsic flaw of plain-text methods, by the way. It is true no matter how you choose to organize your project.) In general, it is worth trying to keep your notes and analysis in a standardized and simple format. The final outputs of projects (such as journal articles or books) tend, as they approach completion, to descend into a rush of specific fixes and adjustments, all running against the ideal of a fully portable, reproducible analysis. It is worth trying to minimize the scope of the inevitable final scramble.

Project organization

Managing projects is a large topic of its own, about which many people have strong opinions. Your goal should be to make your code and data portable, reproducible, and self-contained. To accomplish this, use a project-based approach in RStudio. When you start an analysis with some new data, create a new project containing the data and the R or RMarkdown code you will be working with. It should then be possible, in the ideal case, to move that folder to another computer that also has R, RStudio, and any required libraries installed, and successfully rerun the contents of the project.

In practice that means two things. First, even though R is an object-oriented language, the only "real," persistent things in your project should be the raw data files you start with and the code that operates on them. The code is what is real. Your code manipulates the data and creates all the objects and outputs you need. It's possible to save objects in R, but in general you should not need to do this for everyday analysis.

Second, your code should not refer to any file locations outside of the project folder. The project folder should be the "root" or ground floor for the files inside it. This means you should not use *absolute* file paths to save or refer to data or figures. Instead, use only *relative* paths. A relative path will start at the root of the project. So, for example, you should not load data with a command like this:

```
## An absolute file path.  Notice the leading '/' that starts
## at the very top of the computer's file hierarchy.
my_data <- read_csv("/Users/kjhealy/projects/gss/data/gss.csv")
```

Instead, because you have an R project file started in the gss folder, you can use the here() library to specify a relative path, like this:

```
my_data <- read_csv(here("data", "gss.csv"))
```

While you could type the relative paths out yourself, using here() has the advantage that it will work if, for example, you use Mac OS and you send your project to someone working on Windows. The same rule goes for saving your work, as we saw at the end of chapter 3, when you save individual plots as PDF or PNG files.

Within your project folder, a little organization goes a long way. You should get in the habit of keeping different parts of the project in different subfolders of your working directory (fig. A.9). More complex projects may have a more complex structure, but you can go a long way with some simple organization. RMarkdown files can be in the top level of your working directory, with separate

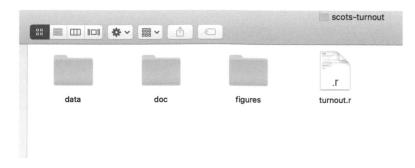

Figure A.9: Folder organization for a simple project.

subfolders called data/ (for your CSV files), one for figures/ (that you might save), and perhaps one called docs/ for information about your project or data files. Rstudio can help with organization as well through its project management features.

Keeping your project organized will prevent you from ending up with huge numbers of files of different kinds all sitting at the top of your working directory.

4 Some Features of This Book

Preparing the county-level maps

The U.S. county-level maps in the socviz library were prepared using shapefiles from the U.S. Census Bureau that were converted to GeoJSON format by Eric Celeste. The code to prepare the imported shapefile was written by Bob Rudis and draws on the rgdal library to do the heavy lifting of importing the shapefile and transforming the projection. Bob's code extracts the (county-identifying) rownames from the imported spatial data frame and then moves Alaska and Hawaii to new locations in the bottom left of the map area so that we can map all fifty states.

First we read in the map file, set the projection, and set up an identifying variable we can work with later on to merge in data. The call to CRS() is a single long line of text conforming to a technical GIS specification defining the projection and other details that the map is encoded in. Long lines of code are conventionally indicated by the backslash character, "\," when we have to artificially break them on the page. Do not type the backslash if you write out this code yourself. We assume the mapfile is named gz_2010_us_050_00_5m.json and is in the data/geojson subfolder of the project directory.

eric.clst.org/Stuff/USGeoJSON

```
# You will need to use install.packages() to install
# these map and GIS libraries if you do not already
# have them.

library(maptools)
library(mapproj)
library(rgeos)
library(rgdal)

us_counties <- readOGR(dsn="data/geojson/gz_2010_us_050_00_5m.json",
                       layer="OGRGeoJSON")

us_counties_aea <- spTransform(us_counties,
                   CRS("+proj=laea +lat_0=45 +lon_0=-100 \
                        +x_0=0 +y_0=0 +a=6370997 +b=6370997 \
                        +units=m +no_defs"))

us_counties_aea@data$id <- rownames(us_counties_aea@data)
```

With the file imported, we then extract, rotate, shrink, and move Alaska, resetting the projection in the process. We also move Hawaii. The areas are identified by their state FIPS codes. We remove the old states and put the new ones back in, and remove Puerto Rico as our examples lack data for this region. If you have data for the area, you can move it between Texas and Florida.

```
alaska <- us_counties_aea[us_counties_aea$STATE == "02", ]
alaska <- elide(alaska, rotate = -50)
alaska <- elide(alaska, scale = max(apply(bbox(alaska), 1, diff))/2.3)
alaska <- elide(alaska, shift = c(-2100000, -2500000))
proj4string(alaska) <- proj4string(us_counties_aea)

hawaii <- us_counties_aea[us_counties_aea$STATE == "15", ]
hawaii <- elide(hawaii, rotate = -35)
hawaii <- elide(hawaii, shift = c(5400000, -1400000))
proj4string(hawaii) <- proj4string(us_counties_aea)

us_counties_aea <- us_counties_aea[!us_counties_aea$STATE %in%
    c("02", "15", "72"), ]
us_counties_aea <- rbind(us_counties_aea, alaska, hawaii)
```

Finally, we tidy the spatial object into a data frame that ggplot can use and clean up the id label by stripping out a prefix from the string.

```
county_map ← tidy(us_counties_aea, region = "GEO_ID")
county_map$id ← stringr::str_replace(county_map$id,
                                pattern = "0500000US", replacement = "")
```

At this point the county_map object is ready to be merged with a table of FIPS-coded U.S. county data using either merge() or left_join(). While I show these steps in detail here, they are also conveniently wrapped as functions in the tidycensus library.

For more detail and code for the merge, see github.com/kjhealy/us-county

This book's plot theme, and its map theme

The ggplot theme used in this book is derived principally from the work (again) of Bob Rudis. His hrbrthemes package provides theme_ipsum(), a compact theme that can be used with the Arial typeface or, in a variant, the freely available Roboto Condensed typeface. The main difference between the theme_book() used here and Rudis's theme_ipsum() is the choice of typeface. The hrbrthemes package can be installed from GitHub in the usual way:

```
devtools::install_github("hrbrmstr/hrbrthemes")
```

The book's theme is also available on GitHub. This package does not include the font files themselves. These are available from Adobe, which makes the typeface.

github.com/kjhealy/myriad

When drawing maps we also used a theme_map() function. This theme begins with the built-in theme_bw() and turns off most of the guide, scale, and panel content that is not needed when presenting a map. It is available through the socviz library. The code looks like this:

```
theme_map ← function(base_size=9, base_family="") {
    require(grid)
    theme_bw(base_size=base_size, base_family=base_family) %+replace%
        theme(axis.line=element_blank(),
                axis.text=element_blank(),
```

```
                axis.ticks=element_blank(),
                axis.title=element_blank(),
                panel.background=element_blank(),
                panel.border=element_blank(),
                panel.grid=element_blank(),
                panel.spacing=unit(0, "lines"),
                plot.background=element_blank(),
                legend.justification = c(0,0),
                legend.position = c(0,0)
                )
}
```

Themes are functions. Creating a theme means writing a function with a sequence of instructions about what thematic elements to modify, and how. We give it a default base_size argument and an empty base_family argument (for the font family). The %+replace% operator in the code is new to us. This is a convenience operator defined by ggplot and used for updating theme elements in bulk. Throughout the book we saw repeated use of the + operator to incrementally add to or tweak the content of a theme, as when we would do + theme(legend.position = "top"). Using + added the instruction to the theme, adjusting whatever was specified and leaving everything else as it was. The %+replace% operator does something similar but has a stronger effect. We begin with theme_bw() and then use a theme() statement to add new content, as usual. The %+replace% operator replaces the entire element specified, rather than adding to it. Any element not specified in the theme() statement will be deleted from the new theme. So this is a way to create themes by starting from existing ones, specifying new elements, and deleting anything not explicitly mentioned. See the documentation for theme_get() for more details. In the function here, you can see each of the thematic elements that are switched off using the element_blank() function.

References

Adelson, E. (1995). *Checkershadow illusion*. Perceptual Science Group @ MIT, http://persci.mit.edu/gallery/checkershadow.

Ai, C., & Norton, E. C. (2003). Interaction terms in logit and probit models. *Economics Letters, 80*, 123–29.

Anderson, E. W., Potter, K. C., Matzen, L. E., Shepherd, J. F., Preston, G. A., & Silva, C. T. (2011). A user study of visualization effectiveness using eeg and cognitive load. In *Proceedings of the 13th Eurographics / IEEE - VGTC Conference on Visualization* (pp. 791–800). Chichester, UK: Eurographs Association; Wiley. https://doi.org/10.1111/j.1467-8659.2011.01928.x

Anscombe, F. (1973). Graphs in statistical analysis. *American Statistician, 27*, 17–21.

Arnold, J. B. (2018). *Ggthemes: Extra themes, scales and geoms for 'ggplot2.'* https://CRAN.R-project.org/package=ggthemes.

Baddeley, A., Turner, R., Rubak, E., Berthelsen Kasper Klitgaard, Cronie Ottmar, Guan Yongtao, Hahn Ute, Jalilian Abdollah, Lieshout Marie-Colette van, McSwiggan Greg, Rajala Thomas, Rakshit Suman, Schuhmacher Dominic, Waagepetersen Rasmus, Adepeju, M., Anderson, C., Ang, Q. W., Austenfeld, M., Azaele, S., Baddeley, M., Beale, C., Bell, M., Bernhardt, R., Bendtsen, T., Bevan, A., Biggerstaff, B., Bilgrau, A., Bischof, L., Biscio, C., Bivand, R., Blanco Moreno, J. M., Bonneu, F., Burgos, J., Byers, S., Chang, Y. M., Chen, J. B., Chernayavsky, I., Chin, Y. C., Christensen B., Coeurjolly J.-F., Colyvas K., Constantine R., Corria Ainslie R., Cotton, R., de la Cruz M., Dalgaard P., D'Antuono M., Das S., Davies T., Diggle P. J., Donnelly P., Dryden I., Eglen S., El-Gabbas A., Fandohan B., Flores O., Ford E. D., Forbes P., Frank S., Franklin J., Funwi-Gabga N., Garcia O., Gault A., Geldmann J., Genton M., Ghalandarayeshi S., Gilbey J., Goldstick J., Grabarnik P., Graf C., Hahn U., Hardegen A., Hansen M. B., Hazelton M., Heikkinen J., Hering M., Herrmann M., Hewson P., Hingee K., Hornik K., Hunziker P., Hywood J., Ihaka R., Icos C., Jammalamadaka A., John-Chandran R., Johnson D., Khanmohammadi M., Klaver R., Kovesi P., Kozmian-Ledward L., Kuhn M., Laake J., Lavancier F., Lawrence T., Lamb R. A., Lee J., Leser G. P., Li H. T., Limitsios G., Lister A., Madin B., Maechler M., Marcus J., Marchikanti K., Mark R., Mateu J., McCullagh P., Mehlig U., Mestre F., Meyer S., Mi X. C., De Middeleer L., Milne R. K., Miranda E., Moller J., Moradi M., Morera Pujol V., Mudrak E., Nair G. M., Najari N., Nava N., Nielsen L. S., Nunes F., Nyengaard J. R., Oehlschlaegel J., Onkelinx T., O'Riordan S., Parilov E., Picka J., Picard N., Porter M., Protsiv S., Raftery A., Rakshit S., Ramage B., Ramon P., Raynaud X., Read N., Reiter M., Renner I., Richardson T. O., Ripley B. D., Rosenbaum E., Rowlingson B., Rudokas J., Rudge J., Ryan C., Safavimanesh F., Sarkka A., Schank C., Schladitz K., Schutte S., Scott B. T., Semboli O., Semecurbe F., Shcherbakov V., Shen G. C., Shi P., Ship H.-J., Silva T. L., Sintorn I.-M., Song Y., Spiess M., Stevenson M., Stucki K., Sumner M., Surovy P., Taylor B., Thorarinsdottir T., Torres L., Turlach B., Tvedebrink T., Ummer K., Uppala M., van Burgel A., Verbeke T.,

Vihtakari M., Villers A., Vinatier F., Voss S., Wagner S., Wang H., Wendrock H., Wild J., Witthoft C., Wong S., Woringer M., Zamboni M. E., Zeileis A., (2017). *Spatstat: Spatial point pattern analysis, model-fitting, simulation, tests.* Retrieved from https://CRAN.R-project.org/package=spatstat.

Bateman, S., Mandryk, R., Gutwin, C., Genest, A., McDine, D., & Brooks, C. (2010). Useful junk? The effects of visual embellishment on comprehension and memorability of charts. In *ACM Conference on Human Factors in Computing Systems (chi 2010)* (pp. 2573–82). Atlanta.

Bates, D., & Maechler, M. (2015). *MatrixModels: Modelling with sparse and dense matrices.* https://CRAN.R-project.org/package=MatrixModels.

———. (2017). *Matrix: Sparse and dense matrix classes and methods.* https://CRAN.R-project.org/package=Matrix.

Bertin, J. (2010). *Semiology of graphics.* Redlands, CA: ESRI Press.

Borkin, M. A., Vo, A. A., Bylinskii, Z., Isola, P., Sunkavalli, S., Oliva, A., & Pfister, H. (2013). What makes a visualization memorable? *IEEE Transactions on Visualization and Computer Graphics (Proceedings of InfoVis 2013).*

Brambor, T., Clark, W., & Golder, M. (2006). Understanding interaction models: Improving empirical analyses. *Political Analysis, 14,* 63–82.

Brownrigg, R. (2017). *Maps: Draw geographical maps.* https://CRAN.R-project.org/package=maps.

Brundson, C., & Comber, L. (2015). *An introduction to R for spatial analysis and mapping.* Londin: Sage.

Bryan, J. (2017). *Gapminder: Data from gapminder.* https://CRAN.R-project.org/package=gapminder.

Cairo, A. (2013). *The functional art: An introduction to information graphics and visualization.* Berkeley, CA: New Riders.

Chakrabarti, R., Haughwout, A., Lee, D., Scally, J., & Klaauw, W. van der. (2017, April). Press briefing on household debt, with a focus on student debt. Federal Reserve Bank of New York.

Chang, W. (2013). *R graphics cookbook.* Sebastopol, CA: O'Reilly Media.

Chatterjee, S., & Firat, A. (2007). Generating data with identical statistics but dissimilar graphics: A follow up to the Anscombe dataset. *American Statistician, 61,* 248–54.

Cleveland, W. S. (1993). *The elements of graphing data.* Summit, NJ: Hobart Press.

———. (1994). *Visualizing data.* Summit, NJ: Hobart Press.

Cleveland, W. S., & McGill, R. (1984). Graphical perception: Theory, experimentation, and application to the development of graphical methods. *Journal of the American Statistical Association, 79,* 531–34.

———. Graphical perception: The visual decoding of quantitative information on graphical displays of data. *Journal of the Royal Statistical Society Series A, 150,* 192–229.

Codd, E. F. (1990). *The relational model for database management: Version 2.* Boston, MA: Addison-Wesley Longman.

Dalgaard, P. (2008). *Introductory statistics with R* (second edition). New York: Springer.

Davies, T. M. (2016). *The book of R.* San Francisco: No Starch Press.

Doherty, M. E., Anderson, R. B., Angott, A. M., & Klopfer, D. S. (2007). The perception of scatterplots. *Perception & Psychophysics, 69,* 1261–72.

Eddelbuettel, D. (2018). *Tint: 'Tint' is not 'Tufte.'* https://CRAN.R-project.org/package=tint.

Few, S. (2009). *Now you see it: Simple visualization techniques for quantitative analysis*. Oakland, CA: Analytics Press.

Fox, J. (2014, December). The rise of the y-axis-zero fundamentalists. https://byjustinfox.com/2014/12/14/the-rise-of-the-y-axis-zero-fundamentalists/.

Freedman Ellis, G. (2017). *Srvyr: 'Dplyr'-like syntax for summary statistics of survey data*. https://CRAN.R-project.org/package=srvyr.

Friendly, M., & Meyer, D. (2017). *Discrete data analysis with R*. Boca Raton, FL: CRC/Chapman; Hall.

Garnier, S. (2017). *Viridis: Default color maps from 'matplotlib' (lite version)*. https://CRAN.R-project.org/package=viridisLite.

———. *Viridis: Default color maps from 'matplotlib'*. https://CRAN.R-project.org/package=viridis.

Gelman, A. (2004). Exploratory data analysis for complex models. *Journal of Computational and Graphical Statistics*, *13*, 755–79.

Gelman, A., & Hill, J. (2018). *Regression and other stories*. New York: Cambridge University Press.

Gould, S. J. (1991). Glow, big glowworm. In *Bully for brontosaurus: Reflections in natural history* (pp. 255–68). New York: Norton.

Harrell, F. (2016). *Regression modeling strategies* (Second edition). New York: Springer.

Healy, K. (2018). *Socviz: Utility functions and data sets for a short course in data visualization*. https://github.com/kjhealy/socviz.

Healy, K., & Moody, J. (2014). Data visualization in sociology. *Annual Review of Sociology*, *40*, 105–28.

Heer, J., & Bostock, M. (2010). Crowdsourcing graphical perception: Using mechanical turk to assess visualization design. In *Proceedings of the Sigchi Conference on Human Factors in Computing Systems* (pp. 203–12). New York: ACM. https://doi.org/10.1145/1753326.1753357.

Henry, L., & Wickham, H. (2017). *Purrr: Functional programming tools*. https://CRAN.R-project.org/package=purrr.

Hewitt, C. (1977). The effect of political democracy and social democracy on equality in industrial societies: A cross-national comparison. *American Sociological Review*, *42*, 450–64.

Imai, K. (2017). *Quantitative social science: An introduction*. Princeton, NJ: Princeton University Press.

Isenberg, P., Bezerianos, A., Dragicevic, P., & Fekete, J.-D. (2011). A study on dual-scale data charts. *IEEE Transactions on Visualization and Computer Graphics*, *17*(12), 2469–87. https://doi.org/10.1109/TVCG.2011.238.

Jackman, R. M. (1980). The impact of outliers on income inequality. *American Sociological Review*, *45*, 344–47.

Koenker, R. (2017). *Quantreg: Quantile regression*. https://CRAN.R-project.org/package=quantreg.

Koenker, R., & Ng, P. (2017). *SparseM: Sparse linear algebra*. https://CRAN.R-project.org/package=SparseM.

Lander, J. P. (2018). *Coefplot: Plots coefficients from fitted models*. https://CRAN.R-project.org/package=coefplot.

Leeper, T. J. (2017). *Margins: Marginal effects for model objects*. https://CRAN.R-project.org/package=margins.

Lumley, T. (2004). Analysis of complex survey samples. *Journal of Statistical Software, Articles*, *9*(8), 1–19. https://doi.org/10.18637/jss.v009.i08.

———. (2010). *Complex surveys: A guide to analysis using R.* New York: Wiley.

———. (2013). *Dichromat: Color schemes for dichromats.* https://CRAN.R -project.org/package=dichromat.

———. (2017). *Survey: Analysis of complex survey samples.* https://CRAN.R -project.org/package=survey.

Matloff, N. (2011). *The art of R programming.* San Francisco: No Starch Press.

Munzer, T. (2014). *Visualization analysis and design.* Boca Raton, FL: CRC Press.

Müller, K. (2017a). *Bindrcpp: An 'rcpp' interface to active bindings.* https://CRAN .R-project.org/package=bindrcpp.

———. (2017b). *Here: A simpler way to find your files.* https://CRAN.R-project .org/package=here.

Müller, K., & Wickham, H. (2018). *Tibble: Simple data frames.* https://CRAN .R-project.org/package=tibble.

Nakayama, K., & Joseph, J. S. (1998). Attention, pattern recognition and popout in visual search. In R. Parasuraman (Ed.), *The attentive brain* (pp. 279–98). Cambridge, MA: MIT Press.

Neuwirth, E. (2014). *RColorBrewer: ColorBrewer palettes.* https://CRAN.R -project.org/package=RColorBrewer.

Openshaw, S. (1983) *The Modifiable Areal Unit Problem.* Norwich: Geo Books.

Pebesma, E. (2018). *Sf: Simple features for R.* https://CRAN.R-project.org /package=sf.

Peng, R. (2016). *programming for data science.* http://leanpub.com/rprogramming.

Pinheiro, J., Bates, D., & R-core. (2017). *Nlme: Linear and nonlinear mixed effects models.* Retrieved from https://CRAN.R-project.org/package=nlme.

Qiu, Y., et al. (2018a). *Sysfonts: Loading fonts into R.* https://CRAN.R-project.org /package=sysfonts.

———. (2018b). *Showtext: Using fonts more easily in R graphs.* https://CRAN .R-project.org/package=showtext.

R Core Team. (2018). *R: A language and environment for statistical computing.* Vienna, Austria: R Foundation for Statistical Computing. https://www .R-project.org/.

Rensink, R. A., & Baldridge, G. (2010). The perception of correlation in scatterplots. *Computer Graphics Forum, 29,* 1203–10.

Ripley, B. (2017). *MASS: Support functions and datasets for Venables and Ripley's mass.* https://CRAN.R-project.org/package=MASS.

Robinson, D. (2017). *Broom: Convert statistical analysis objects into tidy data frames.* https://CRAN.R-project.org/package=broom.

Rudis, B. (2015). *Statebins: U.S. State cartogram heatmaps in R; An alternative to choropleth maps for USA states.* https://CRAN.R-project.org/package =statebins.

Ryan, J. A. (2007). *Defaults: Create global function defaults.* https://CRAN.R -project.org/package=Defaults.

Salganik, M. J. (2018). *Bit by bit: Social research in the digital age.* Princeton, NJ: Princeton University Press.

Sarkar, D. (2008). *Lattice: Multivariate data visualization with R.* New York: Springer.

Silge, J., & Robinson, D. (2017). *Text mining with R.* Sebastopol, CA: O'Reilly. Media.

Slowikowski, K. (2017). *Ggrepel: Repulsive text and label geoms for 'ggplot2.'* https://CRAN.R-project.org/package=ggrepel.

Spinu, V., Grolemund, G., & Wickham, H. (2017). *Lubridate: Make dealing with dates a little easier*. https://CRAN.R-project.org/package=lubridate.

Taub, A. (2016). How stable are democracies? "Warning signs are flashing red." *New York Times*.

Therneau, T. M. (2017). *Survival: Survival analysis*. https://CRAN.R-project.org/package=survival.

Therneau, T., & Atkinson, B. (2018). *Rpart: Recursive partitioning and regression trees*. https://CRAN.R-project.org/package=rpart.

Treisman, A., & Gormican, S. (1988). Feature analysis in early vision: Evidence from search asymmetries. *Psychological Review*, *95*, 15–48.

Tufte, E. R. (1978). *Political control of the economy*. Princeton, NJ: Princeton University Press.

———. (1983). *The visual display of quantitative information*. Cheshire, CT: Graphics Press.

———. (1990). *Envisioning information*. Cheshire, CT: Graphics Press.

———. (1997). *Visual explanations: Images and quantities, evidence and narrative*. Cheshire, CT: Graphics Press.

Vanhove, J. (2016, November). What data patterns can lie behind a correlation coefficient? https://janhove.github.io/teaching/2016/11/21/what-correlations-look-like.

Venables, W., & Ripley, B. (2002). *Modern applied statistics with S* (fourth edition). New York: Springer.

Wainer, H. (1984). How to display data badly. *American Statistician*, *38*, 137–47.

Walker, K. (2018). *Analyzing the US Census with R*. Boca Raton, FL: CRC Press.

Ware, C. (2008). *Visual thinking for design*. Waltham, MA: Morgan Kaufman.

———. (2013). *Information visualization: Perception for design* (third edition). Waltham, MA: Morgan Kaufman.

Wehrwein, A. (2017). Plot inspiration via fivethirtyeight. http://www.austinwehrwein.com/data-visualization/plot-inspiration-via-fivethirtyeight/.

Wickham, H. (2014). Tidy data. *Journal of Statistical Software*, *59*(1), 1–23. https://doi.org/10.18637/jss.v059.i10.

———. (2016). *Ggplot2: Elegant graphics for data analysis*. New York: Springer.

———. (2017a). *Stringr: Simple, consistent wrappers for common string operations*. https://CRAN.R-project.org/package=stringr.

———. (2017b). *Testthat: Unit testing for R*. https://CRAN.R-project.org/package=testthat.

———. (2017c). *Tidyverse: Easily install and load the 'tidyverse.'* https://CRAN.R-project.org/package=tidyverse.

Wickham, H., & Chang, W. (2017). *Devtools: Tools to make developing R packages easier*. https://CRAN.R-project.org/package=dcvtools.

———. (2018). *Ggplot2: Create elegant data visualisations using the grammar of graphics*. http://ggplotz.tidyverse.org.

Wickham, H., & Grolemund, G. (2016). *R for data science*. Sebastopbol, CA: O'Reilly Media.

Wickham, H., & Henry, L. (2017). *Tidyr: Easily tidy data with 'spread()' and 'gather()' functions*. https://CRAN.R-project.org/package=tidyr.

Wickham, H., Francois, R., Henry, L., & Müller, K. (2017a). *Dplyr: A grammar of data manipulation*. https://CRAN.R-project.org/package=dplyr.

Wickham, H., Hester, J., & Francois, R. (2017b). *Readr: Read rectangular text data*. https://CRAN.R-project.org/package=readr.

Wilke, C. O. (2017). *Ggridges: Ridgeline plots in 'ggplot2.'* https://CRAN.R-project.org/package=ggridges.

Wilkinson, L. (2005). *The grammar of graphics* (second edition). New York: Springer.

Xie, Y. (2015). *Dynamic documents with r and knitr* (second edition). New York: Chapman; Hall.

———. (2017). *Knitr: A general-purpose package for dynamic report generation in R.* https://yihui.name/knitr/.

Zeileis, A., & Hornik, K. (2006). *Choosing color palettes for statistical graphics* (Research Report Series / Department of Statistics and Mathematics No. 41). Vienna, Austria: WU Vienna University of Economics; Business. http://epub.wu.ac.at/1404/.

Index

Note: functions, packages, and datasets discussed in the text are indexed in `monospace` type.